Inverse Problems in Underwater Acoustics

Springer
New York
Berlin
Heidelberg
Barcelona
Hong Kong
London
Milan
Paris
Singapore
Tokyo

Michael I. Taroudakis
George Makrakis
Editors

Inverse Problems in Underwater Acoustics

With 76 Figures

Library of Congress Cataloging-in-Publication Data
Inverse problems in underwater acoustics / editors Michael I. Taroudakis, George M. Makrakis.
    p. cm.
Includes bibliographical references and index.

1. Underwater acoustics—Mathematics. 2. Inverse problems (Differential equations) I.
    Taroudakis, M. (Michael I.) II. Makrakis, G.
    QC242.2 .I57 2001
    534 .23′01515353—dc21   CIP00-067227

Printed on acid-free paper.

© 2001 Springer-Verlag New York, Inc.

Springer

Michael I. Taroudakis
Department of Mathematics
University of Crete
714 09 Heraklion – Crete, Greece

George N. Makrakis
Applied and Computational Mathematics
    Foundation
Foundation for Research and Technology
711 10 Herkalion, Greece

Library of Congress Cataloging-in-Publication Data
Inverse problems in underwater acoustics / editors Michael I. Taroudakis, George N. Makrakis.
    p.   cm.
    Includes bibliographical references and index.

    1. Underwater acoustics—Mathematics.   2. Inverse problems (Differential equations)   I.
Taroudakis, Michael I. II. Makrakis, G.
QC242.2.I58 2001
534´.23´01515353—dc21   00-069237

Printed on acid-free paper.

ISBN 978-1-4419-2920-4

9 8 7 6 5 4 3 2 1

Springer-Verlag New York Berlin Heidelberg
A member of BertelsmannSpringer Science+Business Media GmbH

Preface

This volume summarizes some recent developments in selected applications of inverse problems in underwater acoustics. The chapters of the volume are based on presentations made during a research workshop which was held at the Institute of Applied and Computational Mathematics, FORTH, in Heraklion, Crete on May, 1999, and it was also sponsored by the University of Crete. The objectives of this research workshop were to bring together scientists interested in the theoretical issues of inverse problems arising in science and engineering, with people working with specific applications in underwater acoustics, such as sea-bottom characterization, ocean acoustic tomography, and target recognition.

Nowadays, an impressively large number of researchers are looking at these inverse problems from different perspectives. Theoreticians usually deal with the development of notions of solutions, which are both rigorous and appropriate to apply in the inversions of measured data, mainly studying the conditions of existence, uniqueness, and stability of these solutions. On the other hand, applied scientists involved in real world applications, study the conditions under which existing models can be incorporated in inversion schemes under the restrictions posed by the technological tools.

During several recent conferences and workshops [C1], [C2], [C3], [C4], [C5], [C6], [C7], it became apparent that there is a difficulty in applying rigorous inversion schemes in realistic problems as the latter require much more information than is available in the experiments. Moreover, new aspects appearing in the applied field cannot be worked out without the use of advanced mathematical techniques. Therefore theoreticians and applied scientists should work together in order that the inverse problems are treated in the most efficient way. The time for the organization of the workshop and the publication of this volume is appropriate for the following reasons:

- A great amount of experimental data, related to underwater acoustic applications, have been gathered and processed over the last 15 years.

- The processing of these data has demonstrated the necessity for using advanced inversion schemes; as the traditional methods cannot cope with the complex nature of the existing data.

- As new algorithms (such as those related to matched-field processing) are now successfully used for inverting experimental data, there is need for their evaluation in relation to their mathematical justification.

We believe that this volume contains some of the more interesting recent results concerning ocean acoustic tomography, bottom recognition, and inverse scattering for acoustic waveguides, related to the above-mentioned issues. Both newcomers and researchers in underwater acoustics will find clear presentations of the basic ideas, and useful updated references for some of the most important inverse problems in underwater acoustics.

The structure of the volume is as following. Chapters 1–4 deal with the problem of geoacoustic inversions. Chapter 1 comments on the question of what it is possible to recover from a specific experiment for bottom reconstruction. Chapter 2 presents a method for range-dependent recovery of the bottom properties. Chapter 3 is referred to the application of the concept of ocean acoustic tomography for the same problem, while Chapter 4 presents an alternative optimization method also based on tomographic concepts.

Chapter 5 aims at defining whether correlation of deconvolution is superior in estimating the impulse response of an oceanic acoustic waveguide. Chapter 6 refers to the problem of inverting data recorded at an array of hydrophones, when the array shape is an additional unknown of the inverse problem. Chapter 7 presents aspects of a modal inversion scheme applied in ocean acoustic tomography.

The last three chapters emphasize some mathematical aspects of inverse scattering related to the acoustic waveguide which is the underlying model for the problems treated in the previous chapters. In Chapter 8 an inverse scattering technique with incomplete data is applied for the reconstruction of the shape of an embedded object. Chapter 9 deals with the theoretical aspects of an inverse boundary problem for embedded inhomogeneities. Chapter 10 is a comprehensive survey of mathematical results concerning the inverse problem for the acoustic wave equation in several dimensions. The relation of the tomographic data with the underlying oceanographic models is thoroughly investigated in the Appendix, which contains an article published in Ocean Modeling (Paola Malanotte-Rizzoli, Ed.), Elsevier Science BV, Amsterdam, pp. 97–115 (1996). This article provides the kernel of the ocean acoustic tomography problem as it is traditionally used in data assimilation applications.

References

[C1] O. Diachock, A. Caiti, P. Gerstoft, and H. Schmidt (eds.). *Full Field Inversion Methods in Ocean and Seismo-Acoustics*. Kluwer Academic, Amsterdam, 1995.

[C2] J.S. Papadakis (ed.). *Proceedings of the Third European Conference on Underwater Acoustics*. FORTH-IACM, Hemeklion, 1996.

[C3] R. Zhang and J. Zhou (eds.). *Shallow-Water Acoustics*. China Ocean Press, Beijing, 1997.

[C4] R. Chapman and A. Tolstoy (Guest eds.). Special Issue: Benchmarking geoacoustic inversion methods. *J. Comput. Acoust.*, **6** (1&2), 1998.

[C5] A. Alippi and G.B. Cannelli (eds.). *Proceedings of the Fourth European Conference on Underwater Acoustics*. CNR-IDAC, Rome, 1998.

[C6] Y.-C. Teng, E.-C. Shang, Y.-H. Pao, M. Schultz, and A. Pierce. *Theoretical and Computational Acoustics '97*. World Scientific, Singapore, 1999.

[C7] M. Zakharia, P. Chevret, and P. Dubail (eds.). *Proceedings of the Fifth European Conference on Underwater Acoustics*. European Communities, Luxembourg, 2000.

Contents

Erratum
For

Inverse Problems in
Underwater Acoustics

Michael I. Taroudakis and George Makrakis (Editors)

George N. Makrakis's affiliation is incorrect on page iv. The correct affiliation is:

George N. Makrakis
Institute of Applied and Computational Mathematics
Foundation for Research and Technology—Hellas
711 10 Heraklion—Crete, Greece

1

What Are We Inverting For?

David M.F. Chapman

ABSTRACT The goal of geoacoustic inversion is to estimate environmental characteristics from measured acoustic field values, with the aid of a physically realistic computational acoustic model. As modeled fields can be insensitive to variations in some parameters (or coordinated variations in multiple parameters), precise and unique inversions can be difficult to achieve. However, if the results of the inversion are only required for sonar performance prediction (for one example), it is only the resulting acoustic field in the water that matters, often at long range and within a restricted range of frequency. In this context, perhaps a precise description of the seabed structure is not necessary, and it might be sufficient to imagine a simpler seabed model having the same acoustic effect on the underwater sound field within the range-frequency domain of interest. An "effective seabed" model could be built upon the observation that the near-grazing acoustic reflection at the seabed can be described by one complex-valued quantity: the surface impedance. Analytic expressions are derived for an ideal fluid/solid case showing exactly how the five geoacoustic properties reduce to two real parameters: the real part and the imaginary part of the surface impedance at grazing incidence, which are connected with the loss and phase shift of the reflected wave, respectively. It is also shown that several inversion algorithms applied to a typical benchmark problem show variation in the inverted geoacoustic parameters of the seabed model, while all agree on the imaginary part of the impedance (phase shift).

1.1 Introduction

Much progress has been made in ocean acoustic computation, leading to rapid and accurate "forward" model predictions of underwater sound fields, including the effects of complex oceanographic and marine geophysical structure [J+94]. The speed and accuracy of these computer codes, even for complicated environmental inputs, has made it possible to iteratively compare measured acoustic field data with results from computer models to "invert" for model inputs (environmental parameters, source/receiver geometry, or sometimes both). Such acoustic inversion methods—more properly called parameter estimation methods—are the topic of an increasing number of research papers, conference symposia, and workshops.

While exploring the various models, mismatch cost functions, and search strategies available for acoustic inversion, it is important to bear in mind the goal of the activity: that is, what are we inverting for? In the case of geoacoustic inversion, the inversion techniques divide roughly into two types:

(1) those which estimate the geophysical properties of the seabed as precisely as our models will allow; and

(2) those which estimate parameters of an effective seabed model that is adequate for predicting the acoustic field in the ocean.

Inversions of the first type are necessary to construct a "true" picture of the seabed layering and composition: for example, acoustical prospecting for sub-surface mineral resources. The associated geoacoustic models tend to be stratified multiple-layer models of density, sound speed, possibly shear speed, and the associated attenuation coefficients. The model might be range-dependent, possibly continuous in depth (rarely), or piecewise constant or linear (common).

Inversions of the second type may turn out to be more useful for sonar performance prediction models. In this case the geoacoustic model may be simpler, and not bear much relation to the "real" seabed structure, other than reproducing the acoustic effect of the seabed over the range of grazing angles and range of frequencies of interest. This chapter specifically considers inversions of the second type, although the reader will encounter no discussion of acoustic inversion methods or algorithms. Rather, the theme is the physics of acoustic reflection at the seabed in the context of geoacoustic inversion. Furthermore, the material to be presented is clearly biased toward considerations of long-range acoustic propagation in shallow water. Nevertheless, it is hoped that the points raised will appeal to a broader audience and provide food for thought in other cases.

In short, it is proposed that a multiparameter geoacoustic model of the seabed is unnecessary for some sonar applications, and that the expense of time and effort expended in defining a precise geoacoustic model may be unwarranted, when only the acoustic effect of the seabed is important. Furthermore, it is suggested that, at the near-grazing angles of incidence important for long-range propagation, the acoustic surface impedance (also called the normal impedance) characterizes the acoustic effect of the seabed better than the reflection coefficient, although there is a unique mapping from one to the other. At near-grazing incidence, the surface impedance typically has weak dependence on incidence angle and frequency of the sound wave, and the real and imaginary parts of the surface impedance directly relate to reflection loss and phase shift of the wave, respectively. For a seabed simply modeled as a homogeneous viscoelastic solid half-space, it is shown that the five geoacoustic parameters (density, sound speed, shear speed, compressional wave attenuation, and shear wave attenuation) combine nonlinearly to produce a characteristic low-angle complex impedance value, equivalent to just two real parameters. The ambiguity of the geoacoustic parameters leads directly to the concept of "effective" or "equivalent" seabed models. Even complicated multi-parameter seabeds can be modeled simply, if the significant portion of the acoustic field is propagating at near-grazing angles. Conversely, if geoacoustic inversion experiments are not designed carefully, taking into consideration the physics of acoustic reflection and propagation, it may be difficult to unravel the effect of multiple geoacoustic parameters well enough to uniquely define their values. In effect, the information content of the experimental data must match the independent parameters of the model.

1.2 Range and Angle

In his classic 1970 Rayleigh Silver Medal Address [W71], David Weston discussed range-intensity relations in shallow-water underwater acoustics. He showed that the zones of spherical spreading and cylindrical spreading of energy—involving sound waves at steep angles—are confined to relatively short ranges. If the sound speed in water is c_w and the sound speed in the sediment is c_p an important parameter is the critical angle $\theta_c = \arccos(c_w/c_p)$. Because the bottom loss increases dramatically at grazing angles greater than θ_c, one would expect most of the energy to propagate within the wedge of angles $\pm\theta_c$ at ranges beyond $r_c \approx H/2\theta_c$, where H is the water depth. As range increases further, Weston showed that the wedge angle of significant energy progressively decreases, owing to the combined effect of bottom loss and multiple seabed reflections. In the region $r > 6.8H/a\theta_c^2$ the "beam half-width" of the propagating energy is $\theta_{1/2} = (6.8H/ar)^{1/2}$, where a is the bottom loss rate in dB/radian. (This implies a simple linear approximation to the reflection loss versus grazing angle curve, valid at small angles.) Weston calls this the "mode-stripping" zone, as high-order modes (steeply incident waves) are progressively more attenuated.

Figure 1.1 illustrates this effect. It is a SAFARI [S87], [SJ85] model prediction of the EL-b case from the 1997 Matched Field Inversion Workshop [T+98]. The

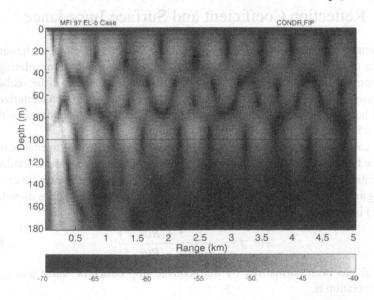

FIGURE 1.1. SAFARI model plot of the acoustic field for the EL-b benchmark inversion case at 40 Hz. Vertical exaggeration is $18\times$. (Plot courtesy of Dave Thomson.)

environment is stratified, with 100 m of water, 75.6 m of homogeneous elastic sediment, and a homogeneous elastic half-space. Note that the 40 Hz acoustic field at short ranges interacts strongly with layers deep beneath the seabed, owing to the steeply propagating waves in this region. As range increases, these steep waves are eventually attenuated, leaving only the near-grazing waves, which sense only the upper seabed layers. We will see that the bottom loss rate for this seabed is about 0.9 dB/radian and the critical grazing angle is 31°. According to Weston's formulas, cylindrical spreading in the water layer persists out to a range of about 2.7 km, after which mode stripping sets in. By a 5 km range, the angular width of the sound field is ±22°. For waves incident at grazing angles below the critical angle, the "transmitted" field in the seabed is actually evanescent in depth, penetrating not much more than 20 m. At a 5 km range, the acoustic field in the ocean does not sense the deep seabed structure; accordingly, an acoustic inversion experiment conducted at this range would be most sensitive to the properties of the sediment, and would be least sensitive to the half-space properties. The acoustic "skin depth" of the seabed scales inversely with frequency, naturally becoming thinner as wavelength becomes smaller. A.O. Williams, in his considerations on the same topic, introduced the concept of "acceptable ignorance" of the seabed below a certain depth [W76].

In summary, for long-range acoustic propagation in shallow water (i.e., many water depths), it is evident that near-grazing angles of propagation dominate the field, and that the near-grazing reflection properties of the seabed are paramount. In turn, the near-surficial sediments prescribe these properties.

1.3   Reflection Coefficient and Surface Impedance

It is common in underwater acoustics to describe the acoustic effect of the seabed in terms of the plane-wave reflection coefficient. However, when considering the reflection of near-grazing sound energy, the surface impedance of the seabed—otherwise known as the normal or input impedance—may better characterize the seabed acoustic properties, as the reflection coefficient varies rapidly over these angles while the impedance is relatively constant.

The acoustic surface impedance of the seabed is identical to the impedance of the acoustic field in the water evaluated at the bottom, that is, the pressure divided by the normal particle velocity. If the reflection coefficient is the arbitrary function of grazing angle and frequency $R(\theta, f)$, it is easy to show that the surface impedance $Z(\theta, f)$ is given by

$$Z(\theta, f) = \frac{1}{\sin \theta} \frac{1 + R(\theta, f)}{1 - R(\theta, f)}, \tag{1.1}$$

where Z has been normalized by the acoustic impedance of water $p_w c_w$. The inverse relation is

$$R(\theta, f) = \frac{Z(\theta, f) \sin \theta - 1}{Z(\theta, f) \sin \theta + 1}. \tag{1.2}$$

Papadakis et al. [P+92] discussed similar relations in connection with an impedance boundary condition for the Parabolic Equation (PE) model of underwater sound propagation. In airborne sound applications, it is common to model soft ground and other locally reacting surfaces by an angle-independent Z that may depend upon frequency [A+95], [E+76], [E96].

It will be shown (at least for an ideal case) that Z approaches a limiting value Z_0 as $\theta \to 0$, and is approximately constant over a wide range of angles. Over the same range of angles R varies linearly with angle, both in amplitude and in phase. For near-grazing angles, this implies that

$$R(\theta) \approx -1 + 2Z_0\theta \qquad (\theta \ll 1). \tag{1.3}$$

It is convenient to decompose Z_0 into real and imaginary components, that is, $Z_0 = X_0 + iY_0$. In polar form, $R(\theta) = |R(\theta)| \exp[i\,\Phi(\theta)]$, and the near-grazing relations between $|R(\theta)|$, $\Phi(\theta)$, X_0, and Y_0 are

$$|R(\theta)| \approx 1 - 2X_0\theta \qquad (\theta \ll 1) \tag{1.4a}$$

and

$$\Phi(\theta) \approx -\pi - 2Y_0\theta \qquad (\theta \ll 1). \tag{1.4b}$$

At near-grazing incidence, the real part of impedance governs the reflection loss and the imaginary part of impedance governs the phase shift. Actually, X_0 is related to Weston's a, as the reflection loss rate turns out to be $17.4X_0$ dB/radian. The quantity Y_0 is related to the effective depth of the seabed [C+89], that is, the phase shift Φ is equivalent to that introduced by an imaginary ideal pressure–release boundary located a distance of $-Y_0/2\pi$ wavelengths below the true boundary. The complex effective depth introduced by Zhang and Tindle is $iZ_0/2\pi$ [ZT93]. A closely linked concept is Joseph's complex reflection phase gradient [J98], which he chooses as an inversion goal in place of the usual geoacoustic parameter set.

1.4 Reflection and Impedance of a Simple Seabed Model

For the purpose of illustration, consider a homogeneous viscoelastic half-space of density ρ, sound speed c_p, and shear speed c_s underneath a water layer of density ρ_w and sound speed c_w. Attenuation of sound waves and shear waves in the seabed are introduced by allowing their speeds to become complex, $c_p \to c_p(1 - i\varepsilon_p)$ and $c_s \to c_s(1 - i\varepsilon_s)$ with ε_p and ε_s small dimensionless numbers.[1] Attenuation of shear waves is usually negligible in this context, but can become extremely important in layered seabeds [CC93], [H+90], [H+91]. The well-known acoustic plane-wave reflection coefficient R is independent of frequency and can be written

---

[1] Multiply ε_p and ε_s by 54.6 to express attenuations in dB/wavelength.

in the form:

$$R(\theta) = \frac{\rho P(\theta)\sin\theta - i\sqrt{\sin^2\theta_c - \sin^2\theta}}{\rho P(\theta)\sin\theta + i\sqrt{\sin^2\theta_c - \sin^2\theta}}, \tag{1.5}$$

in which θ is the grazing angle and $P(\theta)$ is the expression [EC85], [TZ92]:

$$P(\theta) = (1 - 2c_s^2\cos^2\theta)^2 + 4ic_s^3\cos^2\theta\sqrt{1 - c_s^2\cos^2\theta}\sqrt{\sin^2\theta_c - \sin^2\theta}, \tag{1.6}$$

in which we have normalized density by ρ_w and speeds by c_w. All shear wave effects are contained in P, which is unity at normal incidence, and also in the limit $c_s \to 0$. The surface impedance (normalized by $\rho_w c_w$) is then

$$Z(\theta) = \frac{-i\rho P(\theta)}{\sqrt{\sin^2\theta_c - \sin^2\theta}}. \tag{1.7}$$

Brekhovskikh [B60] previously derived this relation in a slightly different form (Equation 27.37). Moreover, §34.2 of [B60] shows how to generalize the result for continuously stratified fluid media.[2]

It is straightforward to derive explicit expressions for X_0 and Y_0 in the case of a viscoelastic solid with $c_s < c_w$. We regard ε_s and ε_p as small quantities and expand, ignoring terms of order ε_p^2, ε_s^2, $\varepsilon_p c_s^3$, $\varepsilon_s c_s^3$, and higher, to get

$$X_0 = \rho\varepsilon_p(1 - 2c_s^2)^2\frac{\cos^2\theta_c}{\sin^3\theta_c} + 4\rho c_s^3\sqrt{1 - c_s^2} + 8\rho\varepsilon_s c_s^2\frac{1 - 2c_s^2}{\sin\theta_c} \tag{1.8a}$$

and

$$Y_0 = -\rho\frac{(1 - 2c_s^2)^2}{\sin\theta_c}. \tag{1.8b}$$

Many observations can be made: The quantity X_0, pertaining to near-grazing reflection loss, is the sum of a compressional-wave-attenuation component, a shear-wave-generation component, and a shear-wave-attenuation component. Substitution of parameter values for typical seabeds [J+94] verifies that the last term in 1.8a is the least significant. The loss due to shear-wave-generation scales as c_s cubed [WE62], [H80]. As sound speed (and critical angle) increases, the loss due to compressional-wave-attenuation decreases. The quantity Y_0, pertaining to a near-grazing phase shift, shows a combination of density and shear speed which is reminiscent of the effective depth formula incorporating shear-wave effects [C+89]. Both X_0 and Y_0 are linear in density: this may seem unremarkable, but is a cautionary note to those who attempt to model seabed reflection effects using models without density discontinuities. Ainslie [A92], [A94] collected several other linear-angle loss expressions, including reflection at a fluid/air boundary and the case of the "hard" seabed, with $c_s > c_w$.

Most importantly, these expressions show precisely how the five viscoelastic seabed parameters map into the two relevant impedance values. In acoustic propagation conditions, where only X_0 and Y_0 determine the field, there are many

---

[2]This material was not retained in the second edition [B80].

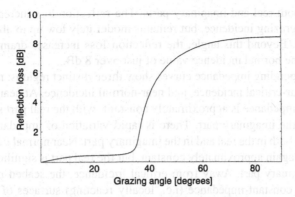

FIGURE 1.2. Reflection loss for a homogeneous seabed of coarse sand.

different seabeds (many families of five parameters) that deliver the same acoustic effect. Conversely, long-range acoustic inversion experiments that effectively only measure X_0 and Y_0 cannot hope to produce unambiguous values for five seabed parameters. However, the above equations strongly suggest how estimated parameter values might be correlated. For example, an inversion algorithm that underestimates shear speed—or perhaps assumes zero shear speed—would naturally also underestimate the density and overestimate the compressional-wave-attenuation, to compensate. Of course, the ambiguity of these inversions could be resolved by sampling the field at short range, where the acoustic field contains information about seabed interaction at steeper angles.

It is illustrative to compare curves of $|R|$ versus θ and Z versus θ for a typical simple case. Consider a seabed of coarse sand, whose parameters might be $\rho/\rho_w = 1.8$, $c_w = 1500$ m/s, $c_p = 1850$ m/s, $c_s = 300$ m/s, $\varepsilon_p = 0.011$ (0.6 dB/wavelength), and $\varepsilon_s = 0.028$ (1.5 dB/wavelength). Figure 1.2 shows the reflection loss curve for this seabed and Figure 1.3 shows the normalized surface

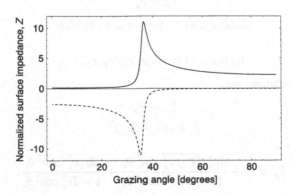

FIGURE 1.3. Surface impedance of a homogeneous seabed of coarse sand. The real part is the solid line; the imaginary part, the dashed line.

impedance, both real and imaginary parts. The reflection loss increases linearly from zero at grazing incidence, but remains moderately low up to about 36°, the critical angle. Beyond this angle, the reflection loss increases dramatically, but levels off at the normal incidence value of just over 8 dB.

The corresponding impedance curves show three distinct regions: near-grazing incidence, near-critical incidence, and near-normal incidence. At near-grazing incidence, the impedance is approximately constant, with the real part significantly smaller than the imaginary part. There is rapid variation of impedance near the critical angle, both in the real and in the imaginary part. Near normal incidence, the impedance is again approximately constant, but the real part is significantly larger than the imaginary part. Away from critical incidence, the seabed behaves like two different constant-impedance (i.e., locally reacting) surfaces of normalized impedance $0.136 - 2.61i$ (near grazing) and $2.22 - 0.0244i$ (near normal).

1.5   Equivalent Seabed Models

The preceding discussion leads naturally to the concept of equivalent or effective seabed models: models that have distinctly different geoacoustic parameters yet have the same acoustic effect, as manifested in the reflection coefficient or surface impedance. Clearly, this cannot be accomplished at all angles simultaneously. In general, to define an equivalent seabed, a range of angles must be specified. In the restricted sense of this chapter, "equivalent seabeds" are identical at zero grazing angle.

A fluid seabed model equivalent to an elastic seabed model must make three parameters (ρ, c_p, and ε_p) do the work of five (ρ, c_p, c_s, ε_p, and ε_s). However, at grazing incidence these five combine into only two relevant quantities, X_0 and Y_0, so there are two equations with three unknowns: an extra condition is required. Typically, one likes to leave the critical angle unchanged, so one possible system of equations is

$$\hat{c}_p = c_p, \tag{1.9a}$$

$$\hat{X}_0(\hat{\rho}, \hat{c}_p, \hat{\varepsilon}) = X_0(\rho, c_p, c_s, \varepsilon_p, \varepsilon_s), \tag{1.9b}$$

and

$$\hat{Y}_0(\hat{\rho}, \hat{c}_p, \hat{\varepsilon}) = Y_0(\rho, c_p, c_s, \varepsilon_p, \varepsilon_s), \tag{1.9c}$$

for which the solutions are

$$\hat{c}_p = c_p, \tag{1.10a}$$

$$\hat{\rho} = \rho(1 - 2c_s^2)^2, \tag{1.10b}$$

and

$$\hat{\varepsilon}_p = \varepsilon_p + \frac{4c_s^3\sqrt{1 - c_s^2}}{(1 - 2c_s^2)^2} \frac{\sin^3\theta_c}{\cos^2\theta_c} + \frac{8\varepsilon_s c_s^2}{1 - 2c_s^2} \frac{\sin^2\theta_c}{\cos^2\theta_c}. \tag{1.10c}$$

This is the classic equivalent fluid model of Tindle and Zhang [TZ92], which they later refined [ZT95]. The effect of shear in the elastic model has been approximated

in the fluid model by reducing the density and increasing the compressional attenuation by factors and terms involving the shear speed and attenuation coefficient.
For example, in the case of the coarse sand seabed described in the last section,
the reduced density turns out to be 1.52 and the augmented attenuation coefficient
is $0.011 + 0.011 + 0.005 = 0.027$ (corresponding to 1.49 dB/wavelength).

Matching only X_0 and Y_0 evidently results in an under-determined system of
equations. The choice of another condition gives rise to several different options for
equivalent seabed models. Possible choices include matching higher derivatives of
the impedance curves at grazing incidence and matching the impedance itself at a
second angle of interest (at normal incidence, for example). These choices necessarily yield complex density values, which are not unknown in acoustic modeling
[A+95], [DB70], [YT99], although many models do not admit them. Another application of equivalent fluid models is to approximate locally reacting surfaces in
propagation models [TM94].

Sometimes effective fluid models make themselves known in unexpected ways.
Consider several "blind" inversions of the EL-b model, for which some of the data
are presented in Table 1.1. [T+98], [A+00] The density, sound speed, and shear
speed for the true sediment layer are shown, along with the "reduced" density calculated from 1.10b and the imaginary part of the impedance calculated from 1.8b.
Three inversions using elastic seabed models are shown, along with one using a
fluid seabed model. The scatter of the inverted density values is significant, especially in the case of the fluid model. The variation in the inverted shear speed values
is even larger. However, the "reduced" density values—which are a combination
of density and shear speed—show remarkable consistency. Moreover, the reduced
densities are very close to the density found by the fluid seabed inversion! In effect,
all these inversions found the same value for the imaginary part of the impedance,
which means the phase of the bottom reflection coefficient was correctly modeled.
Unfortunately, in this case the attenuation coefficient was fixed and known, and
there were no inversions performed on this quantity.

1.6 Reflection and Impedance of a Realistic Seabed Model

The utility of the quantity, $Z_0 = X_0 + iY_0$ the surface impedance at grazing incidence, has been demonstrated for the case of a simple homogeneous viscoelastic

TABLE 1.1. Comparison of parameter inversions for EL-b model

Parameter set	ρ/ρ_w	c_p [m/s]	c_s [m/s]	$\hat{\rho}/\rho_w$	Y_0
True	1.88	1697.8	134	1.82	−3.56
Elastic 1	1.89	1698.0	144	1.82	−3.56
Elastic 2	1.93	1698.0	177	1.82	−3.55
Elastic 3	1.87	1698.0	117	1.83	−3.57
Fluid	1.83	1698.0	—	—	−3.58

solid seabed; however, does the concept work for more complicated seabed models? Earlier, we considered a typical realistic shallow water scenario: the EL-b case from the 1997 Matched Field Inversion Workshop [T+98]. Now consider the acoustic reflection coefficient and the surface impedance of the EL-b environment. In this case the water sound speed (at the bottom) is 1460 m/s and the seabed consists of two homogeneous elastic layers: a sediment layer of density ratio 1.883 and sound speed 1698 m/s, and a lower half-space layer of density ratio 2.146 and sound speed 1840 m/s. The shear speeds are 134 m/s and 214 m/s, respectively. The corresponding sound speed attenuation factors are 0.2469 dB/wavelength and 0.1051 dB/wavelength; the shear speed attenuation factors are 0.25 dB/wavelength in both layers.

The reflection loss at 40 Hz in Figure 1.4 shows low loss up to about 31°, the seabed critical angle. Beyond that angle, there is considerable resonant structure associated with the penetration of sound waves into the finite sediment layer. This structure is echoed in the impedance shown in Figure 1.5, but the impedance values below the critical angle are smooth and relatively constant. The acoustic effect of the EL-b seabed at near-grazing angles is that of a locally reacting seabed of constant impedance $Z_0 = 0.0524 - 3.56i$. This corresponds to a "Weston" bottom loss rate of 0.91 dB/radian and an effective boundary depth of 0.57 wavelengths, which at 40 Hz amounts to 21 m.

The frequency dependence of Z is shown in Figure 1.6 at an incidence angle of 2° and in Figure 1.7 at an incidence angle of 35° (just above critical). Note that is remarkably constant over a wide range of frequencies for low angles, but shows resonant structure at the higher angle.

Before concluding, it would be a mistake to expect that all geoacoustic environments could be so easily interpreted in this simple way. In particular, one should be especially mindful of seabeds with thin upper sediment layers that support significant shear wave resonances. In such cases, the surface impedance can become strongly frequency-dependent [CC93], [A95]. In general, considerations of significantly layered seabed models would naturally lead to a multifrequency inversion

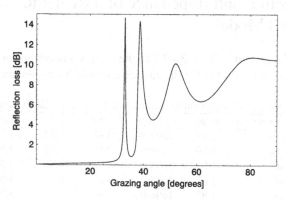

FIGURE 1.4. Reflection loss of the seabed for the EL-b benchmark case at 40 Hz.

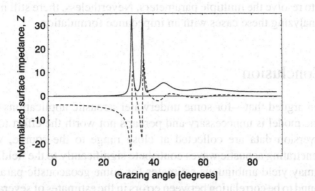

FIGURE 1.5. Surface impedance of the seabed for the EL-b benchmark case at 40 Hz. The real part is the solid line; the imaginary part, the dashed line.

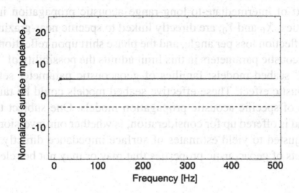

FIGURE 1.6. Frequency dependence of the surface impedance for the EL-b benchmark case at a grazing angle of 2°. The real part (upper curve) is magnified 100×.

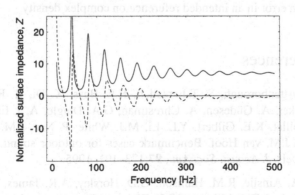

FIGURE 1.7. Frequency dependence of the surface impedance for the EL-b benchmark case at a grazing angle of 35°.

approach to resolve the multiple parameters. Nevertheless, there still may be some merit in analyzing these cases with an impedance formulation.

1.7   Conclusion

It has been argued that—for some underwater acoustic applications—a detailed geoacoustic model is unnecessary and perhaps not worth the effort to determine. Unless inversion data are collected at close range to the source, where steep bottom-penetrating acoustic waves contribute significantly to the field, attempts at inversion may yield ambiguous estimates for some geoacoustic parameters, and there is bound to be correlation between errors in the estimates of several geoacoustic parameters. At near-grazing incidence, the surface impedance Z is relatively constant, and the real and imaginary parts of Z at grazing incidence may represent all that is needed to know (or is knowable?) about the reflectivity of the seabed in the context of intermediate-to-long-range acoustic propagation in the water. These quantities, X_0 and Y_0, are directly linked to specific near-grazing reflection properties: reflection loss per angle, and the phase shift upon reflection. The ambiguity of geoacoustic parameters in this limit admits the possibility of "equivalent" or "effective" seabed models: families of geoacoustic parameter sets that yield the same acoustic effect. These effective seabed models could be tailored to the requirements of specific acoustic propagation models. One subject that was not discussed, and is offered up for consideration, is whether our inversion algorithms should be adjusted to yield estimates of surface impedance directly, rather than ambiguous sets of geoacoustic properties that may or may not be relevant.

Acknowledgments. Thanks to Dave Thomson for providing the SAFARI plot for Figure 1.1. The author also thanks Dale Ellis, Dave Thomson, Mike Ainslie, and Jim Miller for their helpful comments. Thanks to the anonymous reviewer who pointed out an error in an intended reference on complex density.

1.8   References

[A+95]   K. Attenborough, S. Taherzadeh, H.E. Bass, X. Di, R. Raspet, G.R. Becker, A. Güdesen, A. Chrestman, G.A. Daigle, A. L'Espérance, Y. Gabillet, K.E. Gilbert, Y.L. Li, M.J. White, P. Naz, J.M. Noble, and H.A.J.M. van Hoof. Benchmark cases for outdoor sound propagation models. *J. Acoust. Soc. Am.*, **97**:173–191, 1995.

[A+00]   M.A. Ainslie, R.M. Hamson, G.D. Horsley, A.R. James, R.A. Laker, M.A. Lee, D.A. Miles, and S.D. Richards. Deductive multi-tone inversion of sea bed parameters. Accepted by *J. Comput. Acoust.* for inclusion in a special issue on Geoacoustic Inversion, 2000.

[A92]   M.A. Ainslie. The sound pressure field in the ocean due to bottom interacting paths. (Appendix 4) Thesis, University of Southampton, February, 1992.

[A94]   M.A. Ainslie. Shallow water propagation for linear and bilinear sound speed profiles. In *Undersea Defence Technology 94*, pp. 489–493. Nexus, UK, 1994.

[A95]   M.A. Ainslie. Plane-wave reflection and transmission coefficients for a three-layered elastic medium. *J. Acoust. Soc. Am.*, **97**:954–961, 1995.

[B60]   L.M. Brekhovskikh. *Waves in Layered Media*, translated from the Russian by D. Lieberman. Academic Press, New York, 1960.

[B80]   L.M. Brekhovskikh. *Waves in Layered Media*, 2nd ed., translated from the Russian by R.T. Beyer. Academic Press, New York, 1980.

[C⁺89]   D.M.F. Chapman, P.D. Ward, and D.D. Ellis. The effective depth of a Pekeris ocean waveguide, including shear wave effects. *J. Acoust. Soc. Am.*, **85**:648–653, 1989.

[CC93]   N.R. Chapman and D.M.F. Chapman. A coherent ray model of plane-wave reflection from a thin sediment layer. *J. Acoust. Soc. Am.*, **94**:2731–2738, 1993.

[DB70]   M.E. Delaney and E.N. Bazley. Acoustical properties of fibrous absorbent materials. *Appl. Acoust.*, **3**:105–116, 1970.

[E⁺76]   T.F.W. Embleton, J.E. Piercy, and N. Olson. Outdoor sound propagation over ground of finite impedance. *J. Acoust. Soc. Am.*, **59**:267–277, 1976.

[E96]   T.F.W. Embleton. Tutorial on sound propagation outdoors. *J. Acoust. Soc. Am.*, **100**:31–49, 1996.

[EC85]   D.D. Ellis and D.M.F. Chapman. A simple shallow water propagation model including shear wave effects. *J. Acoust. Soc. Am.*, **78**:2087–2095, 1985.

[H⁺90]   S.J. Hughes, D.D. Ellis, D.M.F. Chapman, and P.R. Staal. Low-frequency acoustic propagation in shallow water over hard-rock seabeds covered by a thin layer of elastic-solid sediment. *J. Acoust. Soc. Am.*, **88**:283–297, 1990.

[H⁺91]   S.J. Hughes, D.M.F. Chapman, and N.R. Chapman. The effect of shear wave attenuation on acoustic bottom loss resonance in marine sediments. In J.M. Hovem, M.D. Richardson, and R.D. Stoll (ed.), *Shear Waves in Marine Sediments*, pp. 439–446. Kluwer Academic, Dordrecht, 1991.

[H80]   O.F. Hastrup. Some bottom-reflection anomalies near grazing and their effect on propagation in shallow water. In W. Kuperman and F.B. Jensen (ed.), *Bottom-Interacting Ocean Acoustics*, pp. 135–152. Plenum Press, New York, 1980.

[J+94]   F.B. Jensen, W.A. Kuperman, M.B. Porter, and H. Schmidt. *Computational Ocean Acoustics*, American Institute of Physics, New York, 1994.

[J98]   P. Joseph. Complex reflection phase gradient for the rapid characterization of shallow water sea bottoms. *OCEANS'98 IEEE Conference Proceedings*, pp. 375–379. IEEE, Piscataway, NJ, 1998.

[P+92]   J.S. Papadakis, M.I. Taroudakis, P.J. Papadakis, and B. Mayfield. A new method for a realistic treatment of the sea bottom in the parabolic approximation. *J. Acoust. Soc. Am.*, **92**:2030–2038, 1992.

[S87]   H. Schmidt. SAFARI—Seismo-Acoustic Fast field Algorithm for Range-Independent environments—User's Guide. SACLANTCEN Report SR-113, SACLANT Undersea Research Center, La Spezia, Italy, 15 May, 1987.

[SJ85]   H. Schmidt and F.B. Jensen. Efficient numerical solution technique for wave propagation in horizontally stratified environments. *Comp. Math. Appl.*, **11**:699–715, 1985.

[T+98]   A. Tolstoy, N.R. Chapman, and G. Brooke. Workshop '97: Benchmarking for geoacoustic inversion in shallow water. *J. Comput. Acoust.*, **6**:1–28, 1998.

[TM94]   D.J. Thomson and M.E. Mayfield. An exact radiation condition for use with the a posteriori method. *J. Comput. Acoust.* **2**:113–132, 1994.

[TZ92]   C.T. Tindle and Z.Y. Zhang. An equivalent fluid approximation for a low shear speed ocean bottom. *J. Acoust. Soc. Am.*, **91**:3248–3256, 1992.

[W71]   D.E. Weston. Intensity-range relations in oceanographic acoustics. *J. Sound Vibration*, **18**:271–287, 1971.

[W76]   A.O. Williams. Hidden depths: Acceptable ignorance about ocean bottoms. *J. Acoust. Soc. Am.*, **59**:1175–1179, 1976.

[WE62]   A.O. Williams, Jr., and R.K. Eby. Acoustic attenuation in a liquid layer over a "slow" viscoelastic solid. *J. Acoust. Soc. Am.*, **34**:836–843, 1962.

[YT99]   D. Yevick and D.J. Thomson. Nonlocal boundary conditions for finite-difference parabolic equation solvers. *J. Acoust. Soc. Am.*, **106**:143–150, 1999.

[ZT93]   Z.Y. Zhang and C.T. Tindle. Complex effective depth of the ocean bottom. *J. Acoust. Soc. Am.*, **93**:205–213, 1993.

[ZT95]   Z.Y. Zhang and C.T. Tindle. Improved equivalent fluid approximations for a low shear speed ocean bottom. *J. Acoust. Soc. Am.*, **98**:3391–3396, 1995.

2
Freeze Bath Inversion for Estimation of Geoacoustic Parameters

N. Ross Chapman
Lothar Jaschke

ABSTRACT This chapter describes a new approach for estimating geoacoustic model parameters by matched field inversion of broadband data. The objectives are to design an efficient method to invert accurate estimates of the geoacoustic parameters, and obtain statistical measures of the confidence limits. The essential requirements for the inversion are an efficient search mechanism for exploring the multidimensional model parameter space, and a cost function and propagation model that are appropriate for broadband data. The method should also be robust to sources of mismatch in the experiment, such as imprecise knowledge of the experimental geometry or of the geoacoustic model itself. In this approach, the cost function is based on a multifrequency processor that matches the measured waveform with modeled waveforms that are calculated by ray theory. The search process is a statistical freeze bath algorithm that provides a representation of the distribution of models that fit the data well. The efficiency of the search is improved by reparameterizing, using new parameters based on the covariance of the sampled models. The method is applied to synthetic data that simulate the environment of the Haro Strait tomography experiment.

2.1 Introduction

Inversion of acoustic field data using matched field processing is a widely used practice in ocean acoustics to estimate geoacoustic parameters of the ocean bottom. The inverse problem is posed as an optimization problem that makes use of a global search technique to minimize the mismatch between the measured acoustic field data and replica fields that are calculated for specific parameter values of a geoacoustic model of the ocean bottom. The dimension of the geoacoustic model parameter space is generally very large, so various search algorithms such as simulated annealing (SA) [C+92], [LC93], genetic algorithms (GA) [Ger94], and hybrid methods [Ger95], [FD99], have been used in order to improve the efficiency of the search process. Although these methods have been applied with considerable success to generate acceptable geoacoustic models in various shallow water environments, [CL96], [Tol96], [Ger96] it is a much more difficult problem to assign meaningful objective measures of the uncertainty of the estimates. The inversion of geophysical data is always subject to errors that arise due to experimental errors, modeling errors, or a combination of both. In practice, many models may fit the data well, and the true environmental model is not necessarily the one that yields the optimum fit. Thus, the need to characterize the uncertainty of the estimate is as fundamentally important as the objective of obtaining the estimate itself.

An alternative approach to the optimization method is provided by the statistical formulation of Tarantola [Tar87]. In this approach the solution to the inverse problem is given in terms of the a posteriori distribution or (normalized) probability density in the model parameter space. However, a complete specification of the a posteriori probability density is generally not possible in the multidimensional model parameter space, and one must resort to practical methods of sampling the distribution to determine measures such as the moments or the marginal density functions. The calculation of these measures requires an efficient means of drawing samples from the a posteriori distribution. Various techniques such as Monte Carlo sampling [CC88], [MT95] or Gibbs sampling [SS96] using either SA or GA have been proposed for applications in geophysics. In ocean acoustics, Gerstoft and Mecklenbrauker [GM98] have described a method using GA to generate a posteriori distributions based on the likelihood estimator to estimate geoacoustic parameters and their uncertainty from single frequency and multitone data.

In this chapter we present a statistical freeze bath method for obtaining a representation of the a posteriori distribution of models. The approach is similar to simulated annealing optimization as described by Basu and Frazer [BF90] in that we search for sets of models that optimize a cost (or energy) function that characterizes the misfit between the observed and theoretically predicted data. However, unlike SA which simulates the cooling of an initially high-temperature melt to a final lowest-energy state, the freeze bath samples the model space at constant temperatures that correspond to a uniform freeze probability for all the parameters. An optimally performing freeze bath creates a set of models that consists of a high percentage of different low-energy models that fit the data well. Moreover, the control parameters have intuitive interpretations, and the method itself is straightforward to apply. Contrary to SA optimization, the freeze bath method emphasizes the search for various different low-energy models rather than the search for the best model itself. This behavior makes the freeze bath method especially suitable for inversion in real geophysical problems.

The freeze bath method implemented here was developed to invert broadband acoustic field data from the Haro Strait geoacoustic tomography experiment [C+97]. A more detailed background for matched field inversion for geoacoustic parameters is given in Sections 2.1 and 2.2. The freeze bath method is described in Section 2.3 and is demonstrated in Section 2.4 with simulated data for the Haro Strait environment. The cost function for the search process is based on the correlation of measured and modeled broadband waveforms corresponding to the signals that were received at a vertical line array (VLA) from the light bulb sound sources that were used in the experiment [C+97]. Ray theory is used to model the acoustic propagation to the VLA. Although the environmental model is very simple, it is adequate to reveal basic problems due to parameter sensitivities and correlations that are encountered in geoacoustic inversions in shallow water. We show in Section 2.4 that the efficiency of the sampling process is greatly improved by reparameterizing the initial geoacoustic model using a new independent set of parameters determined from the covariance of the sampled models. The results are summarized in the final section.

2.2   Geoacoustic Inversion

For a particular parameterization of the ocean bottom, each geoacoustic model $\mathbf{m}$ can be represented by a particular set of values for the model parameters

$$\{m_i\} = \mathbf{m} \in \mathbf{M} \qquad \text{for} \quad i = 1, \ldots, N_M, \tag{2.1}$$

where N_M denotes the dimension of the model space $\mathbf{M}$. Similarly, the data may be written as

$$\{d_j\} = \mathbf{d} \in \mathbf{D} \qquad \text{for} \quad j = 1, \ldots, N_D, \tag{2.2}$$

representing an element of the N_D-dimensional data space $\mathbf{D}$. Using this terminology, forward and inverse models are regarded as the rules that connect these two spaces:

$$\text{Forward model:} \quad \mathbf{M} \longrightarrow \mathbf{D}, \tag{2.3}$$

$$\text{Inverse model:} \quad \mathbf{D} \longrightarrow \mathbf{M}. \tag{2.4}$$

Even though the forward problem might be well defined and yield a unique solution, the inverse problem usually does not. Often one may find many different models that fit the data comparatively well. Due to this ambiguity, only very infrequently can an inverse problem be solved in such a way to give direct numerical values of the model parameter. More typically, one is forced to make compromises between the information one actually wants and information that can in fact be obtained from a given dataset.

2.2.1   *Parameter Optimization*

The geoacoustic inversion methods based on a matched field processing attempt to find Earth models that explain experimental observations that are represented by acoustic field data obtained using arrays of hydrophones in the water. In this context the inverse problem is closely related to the mathematical problem of optimization. An optimization problem involves finding an optimal value of a function of several variables. For ocean bottom parameter estimation, the function to minimize (optimize) is a cost function $E(\mathbf{m})$ that characterizes the misfit between the observed acoustic field data $\mathbf{d}$ and the synthetic data predicted by the forward model $\mathbf{d}_{th} = f(\mathbf{m})$. E can also be regarded as an energy function for our purposes here. The set of model parameters, that minimizes the energy function, is the solution of the *optimization* problem.

For problems in which the forward theory is linear, or approximately so over some range of **m**, many optimization procedures are closely related to theories for the generalized inverse of a matrix. Unfortunately, most geophysical problems are not linear, and attempting to linearize them requires assumptions that one would prefer to avoid. The energy function can be specified in various ways, depending on what aspects of the data one wants to include into the comparison with the predicted, or modeled data.

2.2.2 The Maximum Likelihood Estimator

If knowledge of the error statistics is available, it is possible to construct an energy function that estimates the most likely parameter values for the given distribution of errors. Energy functions that are constructed using this approach are therefore called maximum likelihood estimators.

Assuming that a Gaussian function conveniently models the error distribution, the conditional probability that the true data array is **d**, given that the model array is **m**, is expressed by the density function

$$\theta(\mathbf{d} \mid \mathbf{m}) \propto \exp\left(-\tfrac{1}{2}[f(\mathbf{m}) - \mathbf{d}]^T C^{-1}[f(\mathbf{m}) - \mathbf{d}]\right). \qquad (2.5)$$

Here, C denotes the data covariance array and $\theta(\mathbf{d} \mid \mathbf{m})$ is called the likelihood function. Clearly, the maximum of θ coincides with the minimum of the energy function

$$E(\mathbf{m}) = \tfrac{1}{2}[f(\mathbf{m}) - \mathbf{d}]^T C^{-1}[f(\mathbf{m}) - \mathbf{d}]. \qquad (2.6)$$

Note that $E(\mathbf{m})$ in (2.6) depends on the data covariance. Although early theories attributed all errors to inaccurate measurements, in fact, in most geoacoustic inversion problems theory errors are more significant than measurement errors. Even if the underlying physics is completely understood, there are usually many sources of mismatch errors that are not accounted for by the forward model. Such errors may be caused by, for instance, lateral heterogeneity of the actual Earth structure, or mislocation of source and receivers.

Due to these additional sources of errors a reasonable measure of the data covariance is usually not available, even though the measured signal-to-noise ratio might be very well known.

2.2.3 Waveform Matching

An energy function for full field inversion of broadband data was suggested by researchers in geophysics [SS91]. This approach involves matching synthetic waveforms $q_i(t)$, referred to as synthetic seismograms, with measured waveforms $p_i(t)$.

Prediction of the synthetic waveforms at each of N_{rec} hydrophones requires knowledge of the source waveform $s(t)$, and the waveguide's impulse response functions $g_i(t)$, $i = 1, \ldots, N_{rec}$. Synthetic waveforms are obtained by convolving the source waveform by the impulse response functions:

$$q_i(t) = s(t) \star g_i(t). \tag{2.7}$$

A measure of misfit is given by the correlation between the measured and synthetic waveforms:

$$E(\mathbf{m}) = \frac{1}{2} \left(1 - \frac{1}{N_{\text{rec}}} \sum_{i=1}^{N_{\text{rec}}} \frac{\int q_i(t) \cdot p_i(t)\, dt}{\sqrt{\sum_{i=1}^{N_{\text{rec}}} \int q_i^2(t)\, dt} \cdot \sqrt{\sum_{i=1}^{N_{\text{rec}}} \int p_i^2(t)\, dt}} \right). \tag{2.8}$$

In this form, E represents the average negative crosscorrelation of measured waveforms and their modeled counterparts. The normalization is chosen to limit the range of E to the interval $[0, +1]$, with $E = 0$ being a perfect waveform match for all pressure time series.

In contrast to the maximum likelihood estimator, the energy function in (2.8) does not require a priori knowledge of the data error. Thus the model that minimizes (2.8) does not correspond to the most likely model, but to the model that yields the highest average correlation between the measured and modeled data.

2.3 Parameter Uncertainty Estimation

In an optimization method, emphasis is put on the search for the global minimum of a misfit function. Generally, no attempt is made to estimate the structure of this minimum, or possible local minima in other regions of the model space. For example, in the classical optimization problem of the traveling salesman the goal is to find the shortest route that connects a number of places to be visited by the salesman. The existence of other routes nearly as short is not of much interest.

In geophysical inversions, however, a meaningful model estimate consists not only of numerical parameter values at the global minimum of a misfit function. In addition, information on parameter values at points other than the global minimum is necessary to estimate the uncertainty of the solution.

Typically, geoacoustic inverse problems suffer from the fundamental limitation that several models may fit the observations very well. This phenomenon is described mathematically as nonuniqueness of the model solution. The main reasons for this behavior are well known (e.g., Menke [Men84]). As mentioned above, an apparent cause is that the properties of the ocean bottom can in reality vary continuously in all spatial directions, and one is faced with the problem of constructing an Earth model from a finite set of measurements. In this case the inverse problem is highly underdetermined and results in many nonunique solutions. Another cause of nonuniqueness is related to the problem of sensitivity of the model to the data.

Thus, in problems of the type considered here the more important question is often what set of models fits the data well, rather than what single model fits the data best. Modern inverse methods attempt to find this set. In contrast to defining a single (optimal) model as the solution of an inverse problem, a distribution of models is regarded as the solution of an inverse problem. As we will show, knowledge of this distribution can provide valuable insight into the physical system itself.

2.3.1 A Posteriori Probability Density

As an example, consider the maximum likelihood estimator of Section 2.4.2. According to the inverse theory of Tarantola [Tar87] as well as to Bayes' approach [SS96], the a posteriori probability distribution $\sigma(\mathbf{m})$ is obtained by the conjunction of the theoretical likelihood function $\theta(\mathbf{m})$ and the a priori probability density function $\rho(\mathbf{m})$:

$$\sigma(\mathbf{m}) \propto \rho(\mathbf{m}) \cdot \theta(\mathbf{d} \mid \mathbf{m}). \tag{2.9}$$

Here, $\rho(\mathbf{m})$ contains all the information about $\mathbf{m}$ in the absence of data. In the case of Gaussian errors, the likelihood function is given by (2.5) and (2.6), respectively. The a posteriori probability function (representing the solution of the inverse problem) is then given by

$$\sigma(\mathbf{m}) \propto \rho(\mathbf{m}) \cdot \exp(-E(\mathbf{m})). \tag{2.10}$$

Clearly the density is highest in regions of the model space that yield the lowest energies. Once σ has been identified as given by (2.10), the answer to the inverse problem is given by the distribution itself.

However, the result in (2.10) is merely a description of the problem. In reality one is faced with the problem of estimating σ in a large multidimensional model space. A simplistic approach of evaluating σ is to use an optimization technique to locate the minimum of the energy function which usually corresponds to the maximum of σ. Assuming that σ can be approximated by a Gaussian around its peak (if there is such a peak), the covariance can be computed from the curvature at its mean; thus the distribution can be described analytically. In general, however, σ may have a complex, multimodal shape and such an approach will not be effective.

Even if the a posteriori distribution were known, it is difficult to display σ directly in a multidimensional space. For that reason, insight into the nature of the distribution is usually gained by its marginal density functions

$$\sigma_i(m_i) = \int dm_1 \int dm_2 \dots \int dm_{i-1} \int dm_{i+1} \dots \int dm_{N_M} \sigma(\mathbf{m}). \tag{2.11}$$

Equivalent to the knowledge of a distribution is the knowledge of its moments, the most important being the mean:

$$\langle \mathbf{m} \rangle = \int \sigma(\mathbf{m}) \, d\mathbf{m}, \tag{2.12}$$

and the model covariance array:

$$C_M = \int [\mathbf{m} - \langle \mathbf{m} \rangle][\mathbf{m} - \langle \mathbf{m} \rangle]^T \sigma(\mathbf{m}) \, d\mathbf{m}. \tag{2.13}$$

2.3.2   Numerical Integration

All the expressions (2.11)–(2.13) require a numerical integration of the form

$$I = \int \xi(\mathbf{m})\sigma(\mathbf{m}) \, d\mathbf{m}, \qquad (2.14)$$

where $\xi(\mathbf{m})$ is some function of $\mathbf{m}$. The most straightforward way to evaluate I numerically utilizes an enumerative scheme, where the model space is sampled on a dense grid and the integrand is evaluated at each of the grid points. Such a procedure is usually not feasible due to the intensive computational effort required.

An alternative approach is suggested by Monte Carlo integration schemes that make use of pseudo-random number generators. The integrand is evaluated at points chosen at random. Straightforward Monte Carlo methods draw samples from a uniform distribution, many of which do not significantly contribute to the integral. Although there are some rules [Rub81], in practice it is difficult to estimate how many function evaluations are required for accurate estimation of the integral.

If the selection of sample points is taken from regions that are the most important, i.e., which contribute the most to the integral, fewer random trials are necessary for an accurate estimation of (2.14). This requires that, unlike the pure Monte Carlo method, the random values of the model parameters must be drawn from a nonuniform distribution. Such a nonuniform distribution cannot, however, be chosen arbitrarily. Numerical integration methods that follow this approach are known as importance sampling techniques. Gerstoft and Mecklenbrauker have applied importance sampling using genetic algorithms to obtain samples of the a posteriori probability distribution in their work [GM98]. In the next section we describe a different approach based on the method of SA to generate a representation of the distribution that contains mostly realizations of low-energy states.

2.4   Freeze Bath Inversion

The freeze bath method is fundamentally related to the process of SA as described by Basu and Frazer [BF90]. In SA, a control parameter analogous to the temperature of a physical system in thermal equilibrium is reduced to simulate the process of annealing to a final optimum state. The basic difference between the two techniques is the manner in which the temperature is controlled during the process. Whereas SA is associated with a stepwise decrease of temperature, the freeze bath samples new models while holding the temperature constant. The set of sampled models is used to generate a distribution of models that fit the data well. Thus, the freeze bath method provides a measure of the uncertainty bounds for a parameter estimate, in addition to an indication of the best estimated value.

In the physical annealing process of a pure system, the temperature at which a physical system of particles changes its state of aggregation from liquid to solid is called the freezing temperature. When the system is brought to equilibrium at the freezing temperature, frequent transitions between states of the liquid phase and the solid phase occur. Simulating this behavior, the freeze bath repeatedly samples models at a constant temperature that gives preference to low-energy models, but also allows transitions into regions of higher energy. In analogy with the thermodynamic process, the temperature at which the freeze bath operates is defined as the freezing temperature T_*. We can describe the freezing temperature intuitively as the temperature at which the sampling of a parameter is characterized by a general preference of low-energy states but with an adequate probability to escape from energy minima.

2.4.1 Sampling on a Fuzzy Grid

The algorithm used to sample the model parameter space is the heat bath algorithm [Rot85] of SA. The typical heat bath algorithm is as follows:

(1) Initialize

Choose a random starting state $\mathbf{m} = (m_1, m_2, \ldots, m_{N_M})$ and starting temperature $T = T_{start}$.

(2) Sweep

Let $\{\mu_1, \mu_2, \ldots, \mu_k, \ldots, \mu_K\}$ be values within the allowed search range of m_1. Let $\{E_1, E_2, \ldots, E_k, \ldots, E_K\}$ be the corresponding values of a cost function, $E(\mathbf{m})$, with the other components of $\mathbf{m}$ fixed at their current values. Select the new value of m_1 by sampling from the distribution $P_k \propto \exp\{-E_k/T\}$. Visit the remaining components m_k of $\mathbf{m}$ and update them by the same method.

(3) Cool

Reduce T slightly and repeat the sweep step.

(4) Stop

A subjective abort criterion may be defined. For instance, if the model $\mathbf{m}$ has not changed in the last 50 sweeps then the system is frozen, stop.

Sweeping through the parameter space during a heat bath iteration requires the evaluation of the energy function at a number of possible values for each model component. Traditionally, these values were taken from a regular grid [Rot85], [BF90] within predefined (or a priori) bounds for the maximum and minimum allowed value.

However, the drawback of the regular grid search is the inherent resolution limit that is given by the grid spacing. We present a new method, sweeping on a *fuzzy* grid, that allows model components to attain every possible value without a prohibitive amount of computing time. The allowed range of values for a model component μ is divided into K subranges, as shown in Figure 2.1. Each sweep includes two steps. In the first step, a random value μ_k is generated in each subrange (open circle) and the corresponding energy E_k is calculated. The component that was sampled in the previous sweep, μ_{prev} (black circle), is also included in the set of possible values. In the second step, the new component is chosen by sampling from the distribution $p_k \propto \exp(-E_k/T)$. If the newly sampled value is identical

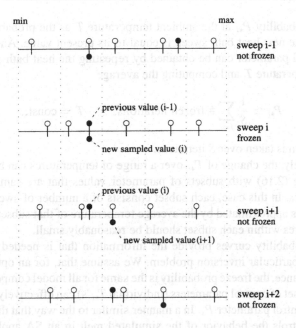

FIGURE 2.1. Sweeping on a fuzzy grid. Frozen iterations are shown in the first two sweeps.

with the previously sampled value, the iteration is said to be frozen. The probability for a frozen iteration to occur depends on the temperature T. In Figure 2.1, the sweep i is an example of a frozen iteration. Since the values $\mu_1 \ldots \mu_K$ are different for each iteration, the overall resolution is not limited to the step size of a regular grid. Each parameter requires only K additional forward model evaluations per sweep, corresponding to the number of subranges; a value of $K = 5$ provided effective sampling and was numerically efficient. The freeze bath inversion proceeds by initially cooling the system down from a high temperature to T_*. Once the cooling process has reached T_*, the ambient temperature is no longer reduced after each sweep, but held constant. Repeated sweeping at T_* produces a set $\{\mathbf{m}\}$ of S models, $\{\mathbf{m}\} = \mathbf{m}_1, \mathbf{m}_2, \ldots, \mathbf{m}_S$, which are distributed like samples drawn from the Boltzmann distribution

$$P(\mathbf{m}) = \frac{\exp\left(-E(\mathbf{m})/T\right)}{\sum \exp\left(-E(\mathbf{m})/T\right)}, \tag{2.15}$$

with $T = T_*$. Clearly, the distribution density is highest for low-energy configurations.

2.4.2 Freeze Probabilities

The freeze bath method requires the ith model component to be sampled at its appropriate freezing temperature T_{*i} or, equivalently, at a temperature that yields a suitable ratio of frozen to nonfrozen heat bath iterations. In order to quantify the tendency for the model component m_i to remain at a constant value, we define its freeze probability P_{*i} at the ambient temperature T as the probability that its value after the next heat bath sweep is equal to its present value. An estimate of this statistical parameter can be obtained by repeating the heat bath algorithm at constant temperature T and computing the average

$$P_{*i} \simeq \frac{1}{S} \sum_{j=1}^{S} \# \text{ frozen iterations}, \qquad T = \text{const.}, \qquad (2.16)$$

where the sum is taken over S iterations.

Alternatively, the change of P_{*i} over a range of temperatures can be estimated by evaluating (2.16) with subsets of parameter values that are sampled during an SA process. In this case, each subset consists of a number of sweeps and the temperature is approximated by the average temperature of that subset. The range of temperatures within each subset should be reasonably small.

Freeze probability curves provide the information that is needed to tune the method to a particular inversion problem. We assume that, for an optimal freeze bath performance, the freeze probability is the same for all model components. This means that a set of control parameters (individual T_{*i}'s) can effectively be reduced to a single control parameter P_*. In a manner similar to the way that the annealing schedule controls the behavior of the simulated melt in an SA application, the freeze probability influences the characteristics of the model set that is sampled by a freeze bath run. P_* represents a trade-off parameter between average energy and model variety of the model set that is sampled during a freeze. Generally, a low-freeze probability will result in a large variety of sampled models with a relatively high-energy average. In contrast to that, a high-freeze probability will yield a lower-energy average, but also a set that lacks variety.

Theoretically, an infinitely long freeze with a uniform sampling *temperature* results in the Boltzmann equilibrium distribution (2.15). By monitoring the energy distribution of the sampled model set {**m**}, its degree of convergence to the Boltzmann distribution can be used to define a stopping criterion. Contrary to that, freezing with uniform *freeze probability* results in an equilibrium distribution different from the Boltzmann distribution. We have not been able to construct the analogous multiple-temperature equilibrium distribution explicitly. However, the particular shape of the limiting energy distribution is only of minor interest. We expect the energy distribution to be a smooth function of energy, with its maximum at low energies. Clearly, one should continue sampling until the energy distribution has converged.

2.4.3 *Reparameterization*

In most applications of geoacoustic inversion there exist correlations between some of the model parameters. These correlations have a significant impact of the efficiency of the search algorithm and the ultimate ability of the freeze bath to determine low-energy models. In the case of a high-correlation coefficient, low-energy models are mostly located along a line that is obliquely oriented to the model coordinate axes, and the parameters are said to be coupled. Since the heat bath algorithm determines a new model by sweeping along lines parallel to the coordinate axes, the navigation along an oblique low-energy valley is very inefficient. This motivates a change of model components m_i to a set of new variables m_i', whose covariance array is diagonal, similar to the approach of Collins and Fishman [CF95]. Such a set is obtained by the orthogonal transformation

$$\mathbf{m}' = \mathbf{A}^T \mathbf{m}, \tag{2.17}$$

where the columns of the transformation matrix $\mathbf{A}$ are identical with the eigenvectors $\mathbf{a}_i$ of the model covariance matrix. The elements m_i' are identified as the coefficients of these eigenvectors. $\mathbf{A}$ can be found by the singular value decomposition of the covariance matrix

$$\text{cov}(\mathbf{m}) = \mathbf{A}\mathbf{\Lambda}\mathbf{A}^T. \tag{2.18}$$

Here, $\mathbf{\Lambda}$ is a diagonal matrix with elements λ_i that are equal to the eigenvalue of the corresponding eigenvector $\mathbf{a}_i$. A representation of the model set in the original coordinate system is obtained by the appropriate inverse transformation

$$\mathbf{m} = \mathbf{A}\mathbf{m}'. \tag{2.19}$$

2.5 Simulation

In order to demonstrate freeze bath inversion, synthetic data were generated to simulate the broadband signals from the Haro Strait geoacoustic tomography experiment [C+97]. The synthetic signals and the replica waveforms were computed by convolving a light bulb waveform recorded in the experiment with the impulse responses calculated using the ray theory code *GAMARAY* [WV87], [WT87]. The inversion is based on matching the broadband waveform with replica waveforms calculated for candidate environments, according to the waveform matching energy function (2.8).

Since the dominant features of the Haro Strait data are the reflections from the sea bottom and a subbottom layer, the environmental model was a simple three-layer system consisting of an isospeed water layer and two solid sediment layers. The compressional speed in the upper sediment layer was inhomogeneous with depth; all other geoacoustic parameters were constant. The model parameter values were taken from Hamilton [Ham80] and are listed in Table 2.1. The environment is assumed to be range independent, and the synthetic data calculated to simulate a light bulb signal for a range of 570 m and source depth of 70 m are shown in Figure 2.2. The data are calculated for receivers spaced at 6.25 m, with the top receiver at 30 m. The bandwidth for the $70 - m$ light bulb is 300 Hz, centered at 600 Hz. The first signal represents the initial bottom reflection and the second signal is the reflection from the lower sediment boundary. The hydrophone numbers refer to the number scheme of the Haro Strait VLA.

TABLE 2.1. Parameter values for the *true* environment.

	Water	Sediment	Half-space
Layer thickness [m]	200.0	40.0	—
c_p at top [m/s]	1482.5	1550.0	1900.0
c_p gradient [1/s]	0.0	2.0	0.0
c_s [m/s]	—	80.0	400.0
ρ [g/cm^3]	1.05	1.7	1.9
k_p [dB/m/kHz]	0.0	0.03	0.03
k_s [dB/m/kHz]	—	1.0	1.0

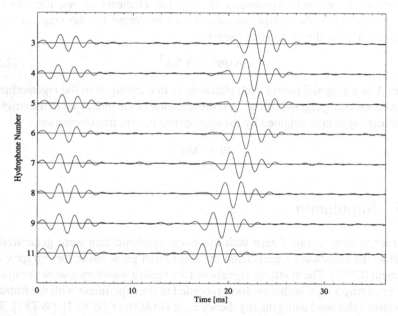

FIGURE 2.2. Synthetic pressure series generated by *GAMARAY* for the environment in Table 2.1.

Previous studies of the sensitivity of the geoacoustic model parameters indicated that the most sensitive parameters were the layer depth, d, and the gradient, c'_L, and compressional speed, c_L, at the top of the upper sediment layer, and the compressional speed of the lower sediment half-space, c_H [C+97]. We have restricted the inversion to estimate values of only these parameters, holding all others at their true values. The parameters and their lower and upper bounds are listed in Table 2.1.

Valuable insight into the structure of the model parameter space can be obtained by investigating the correlations between the parameters. Accordingly, we estimate the correlation coefficient

$$\text{cor}(\mathbf{m})_{ij} = \frac{\text{cov}(\mathbf{m})_{ij}}{\sqrt{\text{cov}(\mathbf{m})_{ii} \cdot \text{cov}(\mathbf{m})_{jj}}}, \tag{2.20}$$

TABLE 2.2. Parameter correlation coefficients.

	d	c_L	c_L'	c_H
d	+1.00	+0.18	+0.23	−0.01
c_L	+0.18	+1.00	−0.63	+0.04
c_L'	+0.23	−0.63	+1.00	−0.10
c_H	−0.01	+0.04	−0.10	+1.00

where

$$\text{cov}(\mathbf{m}) \simeq \frac{1}{S} \sum_{s=1}^{S} [\mathbf{m}_s - \langle \mathbf{m} \rangle][\mathbf{m}_s - \langle \mathbf{m} \rangle]^T, \qquad (2.21)$$

is the model covariance, and

$$\langle \mathbf{m} \rangle \simeq \frac{1}{S} \sum_{s=1}^{S} \mathbf{m}_s \qquad (2.22)$$

is the mean model. Since the model parameters have different units, we calculate (2.22) and (2.21) in normalized model coordinates. The normalization was done by transforming the minimal and maximal parameter bounds to the dimensionless values of 0 and 1, respectively.

The calculated values of cor(**m**) are given in Table 2.2. The highest correlation is observed between the compressional speed gradient and the compressional speed at the top of the layer ($|\text{cor}(\mathbf{m})| = 0.63$) and moderate correlation between these parameters and the layer thickness ($\text{cor}(\mathbf{m}) \simeq 0.2$). The compressional speed in the half-space shows only a small dependence on any of the other parameters. Based on the correlations shown in the table, a new set of four independent parameters was obtained using the orthogonal transformation (2.17) and (2.18).

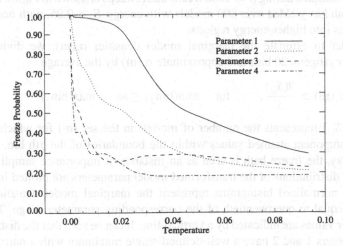

FIGURE 2.3. Freeze probability curves of the rotated parameters $\mathbf{m}' = \mathbf{A}^T \mathbf{m}$.

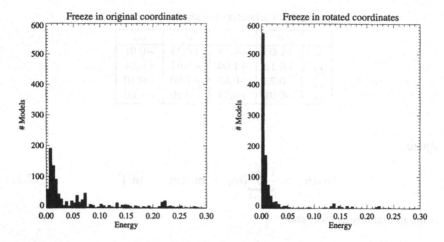

FIGURE 2.4. Histograms of energy values obtained by the freeze bath in original coordinates (left), and obtained by the freeze bath in rotated coordinates (right).

Freeze probability curves estimated following the procedure in Section 2.4.2 are shown in Figure 2.3 for the new set of transformed or rotated parameters. The parameters m'_i are ordered according to eigenvectors $\mathbf{a}_i$ with increasing eigenvalue λ_i. The figure shows that the dominant parameter is the coefficient of the eigenvalue that corresponds to the smallest eigenvalue. In the new parameterization, the dominant components freeze out sequentially, indicating a measure of independence. Parameter sampling temperatures were estimated for a common freeze probability of $P_* \simeq 0.65$. This value of P_* yields a sampling temperature T_* for the most sensitive parameter that is consistent with the temperature at which it begins to freeze in a conventional SA cooling process. The histogram of models that were sampled during $S = 1000$ freeze bath sweeps is shown in Figure 2.4. The freeze bath has yielded over 800 models with energy $E < 0.05$, with occasional excursions into higher-energy regions.

In order to estimate the marginal model densities $\sigma_i(\mathbf{m})$, we divide each parameter range in 20 bins and approximate $\sigma_i(\mathbf{m})$ by the average

$$\sigma_i(\mathbf{m}) \simeq \frac{\#(S_{ij})}{S}, \qquad \text{for} \quad \min(\text{bin}_j) \leq m_i < \max(\text{bin}_j), \qquad (2.23)$$

where $\#(S_{ij})$ represents the number of models in the set $\{\mathbf{m}\}$ for which the ith model component attained values within the boundaries of the jth bin. Viewed in this way, the freeze bath is seen as an instance of importance sampling. The marginal distributions of the transformed model parameters are plotted in Figure 2.5. The normalized histograms represent the marginal model densities, with binsizes equal to one-twentieth of the corresponding parameter range. The true parameter values are indicated by a vertical line. It can be seen that the distribution of parameters 1 and 2 have a well-defined single maximum with a narrow peak close to the true value. The model set in the original geoacoustic parameters is

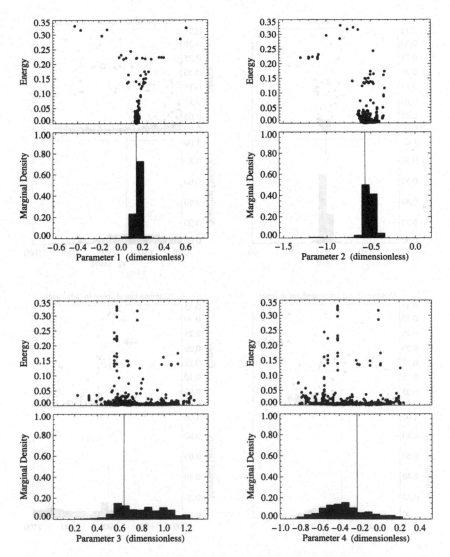

FIGURE 2.5. Distribution of rotated model parameters as a result of heat bath sampling along eigenvectors of the model covariance. The true values are indicated by a vertical line.

obtained by the appropriate inverse transformation

$$\mathbf{m} = \mathbf{A}\mathbf{m}', \qquad (2.24)$$

and the model distributions are shown in Figure 2.6. Although only one parameter (the layer thickness) can be resolved, the freeze was successful in finding low-energy states throughout widespread parts of the search space. In comparison with the distribution of models obtained in a freeze bath in the original parameters (Figure 2.4), the freeze in the rotated parameters has located significantly more

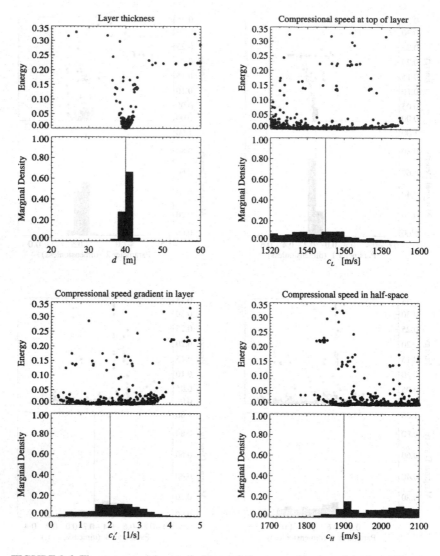

FIGURE 2.6. The same model set as in Figure 2.5 expressed in the original coordinates.

low-energy models with energy $E < 0.05$. This result implies a more efficient sampling process that has produced a faster convergence of the marginal parameter density distributions.

The model set is plotted in two-dimensional subspaces in Figure 2.7. It can be seen that the low-energy valleys were exhaustively sampled (cases (a)–(d)) independent of their orientation with respect to the coordinate axes. Given this thorough sampling, the freeze bath in the rotated coordinates has provided considerable insight into the structure of the parameter space. For instance, the solutions available from a standard SA optimization process are now seen to belong to long connected

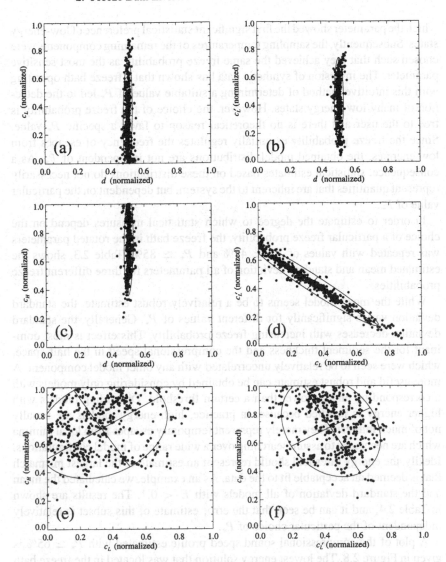

FIGURE 2.7. The original geoacoustic model parameters plotted in normalized, two-dimensional subspaces. The ellipses represent the joint 95% confidence area of the corresponding two-dimensional model distributions. Only models with $E < 0.1$ are considered.

regions of low-energy states, rather than representing individual local minima. If, as in this case, the energy function is unimodal, the mean model is usually a good estimator of the low-energy region.

The model distributions were obtained using the specific freeze probability of $P_* \simeq 65\%$. The choice of this value was based on the freeze probability of the most sensitive parameter, using as a guideline the behavior at temperatures for which the parameter showed the first significant statistical preference of low-energy states. Subsequently, the sampling temperatures of the remaining components were chosen such that they achieved the same freeze probability as the most sensitive parameter. The inversion of synthetic data has shown that a freeze bath operating with this intuitive method of determining a suitable value of P_* led to the detection of many low-energy states. However, the choice of the freeze probability is free to the user and there is no theoretical reason to favor a specific P_* value. Since the freeze probability essentially regulates the frequency of escapes from low energies, the obtained model distributions are not independent of P_*. As a consequence, statistical estimates based on these distributions do not necessarily represent quantities that are inherent to the system, but dependent on the particular value of P_*.

In order to estimate the degree to which statistical measures depend on the choice of a particular freeze probability, the freeze bath in the rotated parameters was repeated with values of $P_* \simeq 45\%$ and $P_* \simeq 85\%$. Table 2.3, shows the estimated mean and standard deviation of all parameters for three different freeze probabilities.

While the mean model seems to be a relatively robust estimate, the standard deviation varies significantly for different values of P_*. Generally, the standard deviation decreases with increasing freeze probability. This effect is most dominant for the sediment thickness and the compressional speed in the half-space, which were seen to be relatively uncorrelated with any other model component. A more useful and robust estimate can be obtained by considering only models with a corresponding energy lower than a certain threshold. Discarding models with higher energy is justified because, in practice, high-energy models are usually not of much interest. They merely represent temporary escapes from local minima which are necessary to assure searching over a wide range of model configurations. Ideally, the energy threshold should represent an estimate of the largest mismatch that is deemed an acceptable fit to the data. As an example, we calculated the mean and the standard deviation of all models with $E < 0.1$. The results are shown in Table 2.4, and it can be seen that the error estimate of this subset is relatively independent of the particular choice of P_*.

A plot of the compressional sound speed profile estimated with $P_* \simeq 65\%$ is given in Figure 2.8. The lowest energy solution that was located in the freeze bath is in excellent agreement with the true values for all the parameters. However, the fact that the true value for the compressional speed in the half-space lies outside

TABLE 2.3. Dependence of the estimated model mean and the standard deviation on the freeze probability.

	d [m]	c_L [m/s]	c'_L [1/s]	c_H [m/s]
$P_* \simeq 45\%$	40.8 ± 4.1	1554 ± 23	2.1 ± 1.1	1936 ± 116
$P_* \simeq 65\%$	40.4 ± 2.0	1547 ± 16	2.2 ± 0.8	1979 ± 73
$P_* \simeq 85\%$	40.2 ± 0.5	1541 ± 18	2.2 ± 0.8	1989 ± 76

TABLE 2.4. Estimated model mean and the standard deviation of all models with $E < 0.1$

	d [m]	c_L [m/s]	c'_L [1/s]	c_H [m/s]
$P_* \simeq 45\%$	40.4 ± 0.8	1553 ± 23	2.0 ± 1.1	1970 ± 81
$P_* \simeq 65\%$	40.2 ± 0.6	1546 ± 16	2.2 ± 0.8	1983 ± 72
$P_* \simeq 85\%$	40.1 ± 0.4	1546 ± 18	2.2 ± 0.8	1989 ± 76

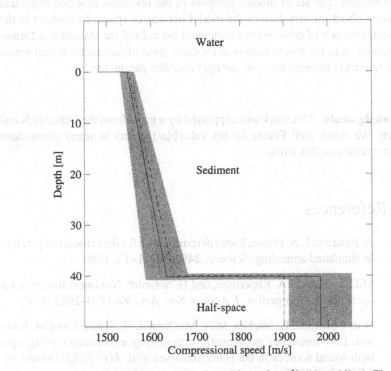

FIGURE 2.8. Best estimated compressional sound speed profile (dotted line). The dashed line represents the true profile, and the continuous line represents the mean of the models with energy $E < 0.1$ (estimated with $P_* \simeq 65\%$). The shaded area covers all parameter values within the corresponding standard deviation.

the standard deviation interval shows that the average of many good models is not necessarily a good model itself.

2.6   Summary

The freeze bath method is a general inversion technique for estimation of geoacoustic model parameters from acoustic field data. Unlike conventional simulated annealing the freeze bath generates a set of models that fit the acoustic data well. In this sense, the method provides a representation of the a posteriori distribution of models that indicates which parameters have been well estimated. We have also introduced an efficient fuzzy grid sampling technique for the conventional heat bath algorithm that improves the resolution of the sampling process.

Freeze bath inversion was demonstrated for a synthetic dataset that simulated the environment of the Haro Strait experiment. The efficiency of the search is improved by transforming to a new parameter set, based on the covariance of the sampled models. The set of models sampled in the inversion revealed important information about the structure of the model parameter space. In contrast to the traditional approach of conventional simulated annealing, the statistical information obtained from the freeze bath enables estimation of parameter uncertainties for both relatively independent and strongly coupled parameters.

Acknowledgments. This work was supported by a grant from the Office of Naval Research. We thank Neil Frazer for his valuable insights in many discussions during the course of this work.

2.7   References

[BF90]   A. Basu and L.N. Frazer. Rapid determination for the critical temperature in simulated annealing. *Science*, **249**:1409–1412, 1990.

[C⁺92]   M.D. Collins, N.A. Kuperman, and H. Schmidt. Nonlinear inversion for ocean bottom properties. *J. Acoust. Soc. Am.*, **92**:2770–2883, 1992.

[C⁺97]   N.R. Chapman, L. Jaschke, M.A. McDonald, H. Schmidt and M. Johnson. Low-frequency geoacoustic tomography experiments using light bulb sound sources in the Haro Strait sea trial. *MTS/IEEE Oceans 97*, **2**:763–768, 1997.

[CC88]   P.W. Cary and C.H. Chapman. Automatic one-dimensional waveform inversion of marine seismic reflection data. *Geophys. J.*, **93**:527–546, 1988.

[CF95]   M.D. Collins and L. Fishman. Efficient navigation of parameter landscapes. *J. Acoust. Soc. Am.*, **98**:1637–1644, 1995.

[CL96]   N.R. Chapman and C.E. Lindsay. Matched field inversion for geoacoustic parameters in shallow water. *IEEE J. Ocean. Eng.*, **21**:347–354, 1996.

[FD99]   M.R. Fallat and S.E. Dosso. Geoacoustic inversion via local, global and hybrid algorithms. *J. Acoust. Soc. Am.*, **105**:3219–3230, 1999.

[Ger94]   P. Gerstoft. Inversion of seismo-acoustic data using genetic algorithms and *a posteriori* probability distributions. *J. Acoust. Soc. Am.*, **95**:770–782, 1994.

[Ger95]   P. Gerstoft. Inversion of acoustic data using a combination of ge-
          netic algorithms and the Gauss–Newton approach. *J. Acoust. Soc. Am.*,
          97:2181–2190, 1995.

[Ger96]   P. Gerstoft and D.F. Gingras. Parameter estimation using multi-
          frequency, range-dependent acoustic data in shallow water. *J. Acoust.
          Soc. Am.*, **99**:2839–2850, 1996.

[GM98]   P. Gerstoft and C.F. Mecklenbrauker. Ocean acoustic inversion with
          estimation of *a posteriori* probability distributions. *J. Acoust. Soc. Am.*,
          104:808–819, 1998.

[Ham80]   E.L. Hamilton. Geoacoustic modeling of the sea floor. *J. Acoust. Soc.
          Am.*, **68**:1313–1340, 1980.

[LC93]   C.E. Lindsay and N.R. Chapman. Matched field inversion for geoa-
          coustic model parameters using adaptive simulated annealing. *IEEE J.
          Ocean. Eng.*, **18**:224–231, 1993.

[Men84]   W. Menke. *Geophysical Data Analysis: Discrete Inverse Theory.*
          Academic Press, New York, 1984.

[MT95]   K. Mosegaard and A. Tarantola. Monte Carlo sampling of solutions to
          inverse problems. *J. Geophys. Res.*, **100**:12431–12447, 1995.

[Rot85]   D.H. Rothman. Nonlinear inversion, statistical mechanics and residual
          static corrections. *Geophysics*, **50**:2784–2796, 1985.

[Rub81]   R.Y. Rubinstein. *Simulation and the Monte Carlo Method.* Wiley, New
          York, 1981.

[SS91]   M.K. Sen and P.L. Stoffa. Nonlinear one-dimensional seismic waveform
          inversion using simulated annealing. *Geophysics*, **56**:1624–1638, 1991.

[SS96]   M.K. Sen and P.L. Stoffa. Bayesian inference, Gibbs sampler and
          uncertainty estimation in geophysical inversion. *Geophys. Prospect.*,
          44:313–350, 1996.

[Tar87]   A. Tarantola. *Inverse Problem Theory: Methods for Data Fitting and
          Model Parameter Estimation.* Elsevier Science, Amsterdam, 1987.

[Tol96]   A. Tolstoy. Using matched field processing to estimate shallow-water
          bottom properties from shot data in the Mediterranean Sea. *IEEE J.
          Ocean. Eng.*, **21**:471–479, 1996.

[WT87]   E.K. Westwood and C.T. Tindle. Shallow-water time series simulation
          using ray theory. *J. Acoust. Soc. Am.*, **81**:1752–1761, 1987.

[WV87]   E.K. Westwood and P.J. Vidmar. Eigenray finding and time series sim-
          ulation in a layered-bottom ocean. *J. Acoust. Soc. Am.*, **81**:912–924,
          1987.

3

Tomographic Inversion on Multiple Receivers/Arrays from Multiple Sources for the Estimation of Shallow Water Bottom Properties

Alex Tolstoy

ABSTRACT Geoacoustic inversion in even rather simplified range independent regions (but for multiple unknown parameters) is known to be quite difficult [TCB]. Additionally, inversion for even *one* parameter but in a range-*dependent* region is also quite difficult [DYOC]. As might be expected, a combination of range variability *plus* multiple parameters leads to an extremely difficult problem. Recent efforts have concentrated on an approach which estimates range-independent, i.e., range-*averaged*, parameters on individual source–receiver/array (SR) paths and then combines all the results in a matrix inversion to estimate the range-*dependent* region properties. This chapter is analogous to one successfully developed for the estimation of large volume ocean sound-speed profiles (as seen in [T3]). This chapter will discuss recent efforts for this new shallow water tomographic geoacoustic inversion via individual paths.

3.1 Introduction

The inversion of acoustic data for the estimation of environmental parameters (such as sediment thicknesses, sound-speeds, densities, and attenuations) is known to be plagued by nonuniqueness issues and by difficulties in navigating enormous solution search spaces. This is the case even for very simple, range-independent scenarios.

In the summer of 1997 an inversion workshop was held in Vancouver, BC, Canada, for the express purpose of studying the behavior of contemporary geoacoustic inversion methods (based on such diverse approaches as simulated annealing, genetic algorithms, broadband iterative methods, neural nets, gradient descent, etc.) as applied to simulated, shallow water, range-independent, acoustic data. A variety of test cases was generated, i.e., six case types (SD, AT, SO, WA, EL, N), each with three subcases (a, b, c) and each varying the unknown parameter values within specified intervals. The test cases increased in difficulty by requiring the estimation of more and more parameters. In Figure 3.1 we see the simplest test case (SD) where only six unknowns were to be estimated:

- sediment thickness h_{sed};

SD
Single sediment layer plus half-space.

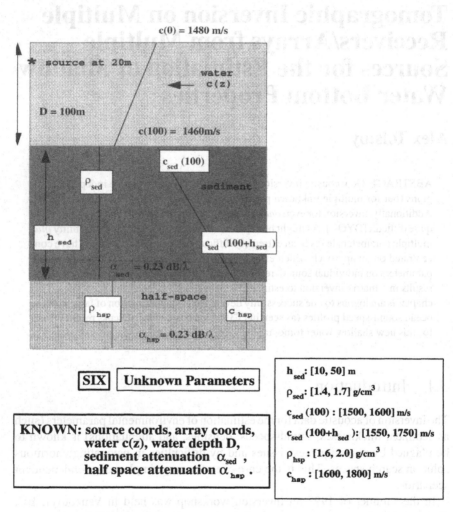

FIGURE 3.1. Diagram for the SD test case of Workshop97.

- constant sediment density ρ_{sed};

- a $1/c^2$ linear sediment sound-speed profile specified by $c_{sed}(100)$ and $c_{sed}(100+h_{sed})$;

- constant half-space sound-speed c_{hsp}; and

- constant half-space density ρ_{hsp};

with corresponding search intervals indicated in the lower right. For the SD sub-cases source range and depth were given exactly, water depth was given as exactly 100 m, the water sound-speed profile was given exactly (a $1/c^2$ linear profile), bottom attenuations were given exactly, and receiver (field) depths were known at all exact 1 m increments. For the five other cases there were different unknown parameters, e.g., for the AT subcases the bottom attenuations also had to be es-timated. Five of the six cases assumed a single (but different for each case and subcase) shallow water sediment layer over a simple half-space; the sixth case N allowed for an unknown number of constant sediment layers. One of the cases, i.e., three of the eighteen subcases (ELa, ELb, and ELc) allowed for elasticity of the sediment and half-space. The simulated no-noise acoustic field data were cal-culated (at 1 Hz increments between 25 and 200 Hz, at 5 Hz increments between 200 and 500 Hz) by SAFARI (a well-documented and highly regarded propagation model deemed suitable for inversion benchmarking at all the low frequencies of interest, i.e., at 25 to 500 Hz; see [JKPS]). All six cases (eighteen subcases) were range-independent (see [TCB], for more details).

It was found that some properties were easier to estimate than others, i.e., were more dominant (such as h_{sed} and $c_{sed}(100)$ relative to ρ_{sed}); the field sensitivities to some properties were correlated (such as $c_{sed}(100)$ with $c_{sed}(100+h_{sed})$); and some methods found it advantageous to use multiple frequencies or iterative appraoches in their processing schemes. In general the Workshop97 results indicated that even simple range-independent scenarios (all noise free) could be *very difficult* to invert for the estimation of bottom properties.

Since Workshop97 there has been a follow-on workshop (SWAM99) held in Monterey, CA, and dedicated to analysis of the performance of range-*dependent* propagation modeling where such modeling will be an integral component of inversion in range-*dependent* scenarios. While there is no single propagation model which is credited as a "benchmark" solution for any given two-dimensional or three-dimensional environment, it is generally recognized that careful use of a coupled normal-mode code (as first suggested and demonstrated in [JF]) should provide an excellent basis for comparison with models such as PE or ray-based codes. SWAM99 offered a number of test cases for the application of propagation codes (more realistic than the ideal wedge or deSanto eddy), thus providing a uniform set of situations followed by a uniform format for the output fields. This allowed for inter-comparisons between models such as PE, adiabatic and normal modes, and ray-based codes, and among newly created models versus "standard" models.

SWAM99 found that even implementation of models by "knowledgeable" (but not necessarily "expert") users could result in a wide disparity of predictions (up to tens of dBs in transmission loss for the fields at selected ranges, depths, and frequencies). That is, running range-*dependent* propagation models is *nontrivial*. Thus, incorporating such models into geoacoustic inversion methods promises to be problematic—the next step for most inversion approaches which hope to be ap-plied to more realistic shallow water situations than those devised for Workshop97. If we find in the meantime that adiabatic normal mode modeling is sufficiently accurate for many of the SWAM99 cases, e.g., for shallow water environments, then it can readily be incorporated into our tomographic method (as discussed in [T5]).

The tomographic method to be discussed in this chapter will involve two steps:

(1) the implementation of a *user selected* range-independent inversion method (such as one of those used for the Workshop97 cases, e.g., the RIGS method discussed in [T2]) applied to each single SR path sampling the region; followed by

(2) a final, single matrix inversion to calculate the fully range-*dependent* environmental parameters.

The method assumes that out-of-plane propagation, i.e., with full three-dimensional horizontal refraction effects, is *not* significant, and it will *not* explicitly account for such effects. Thus, we assume an Nx2D approach. However, general *range* and *azimuthal* dependence *will* be estimated (assuming adiabatic normal mode propagation).

It should be noted that an MFP-based approach was attempted originally which implemented an adiabatic normal mode approach plus inversion for each *single range-varying* bottom parameter (such as sound-speed $h_{sed}(100)$ at the water–sediment interface) along *one* source-to-receiver array (SR) path. This proved to be excessively difficult with results requiring long computation times and producing significant errors in the parameter estimates. Thus, the estimation of even one range-varying parameter along a *single* SR path (as attempted by Dosso et al., [DYOC]) has since been replaced by the tomographic method to be described next.

3.2   Approach

The method to be applied here has its history in deep water Matched Field Processing (MFP; see [T4]) with application to tomographic sound-speed estimation over large volumes (see [T1], [T3]). It requires multiple arrays of receivers (preferably vertical arrays spanning the water column), and a wide distribution of sources generating broadband low-frequency signals (less than a few hundred Hz) to which MFP is applied. The correlation between "data" (so far only simulated data have been considered) and model predictions (where environmental inputs are varied in some systematic manner) is maximized per SR path via MFP. For shallow water geoacoustic inversion this maximization per path may be over a variety of frequencies (often a simple incoherent summing of frequencies is used as in the RIGS method). We assume that the maximation over range-independent estimated geoacoustic parameters *for each single SR path* will occur for the "average" parameters for this path.

The method does assume that adiabatic normal modes are an appropriate description of acoustic propagation behavior (an assumption embedded in the linearization of the problem). While this may be considered a significant restriction by some researchers, it may not be as severe as originally thought since the use of appropriate normal mode modeling may be sufficient to overcome the usual coupling behavior.[1]

The tomographic approach involves:

(1) *Estimating* for each SR path (total of N paths) the *average* per path environmental parameter $\overline{p_n}$ by some inversion method (such as RIGS) where n is the SR path index.

(2) *Gridding* the region into M cells. The number of cells selected is a trade-off between desirable high-resolution range and azimuthal variability to match reality (M large) and available acoustic path sampling ($M \ll N$).

(3) *Inverting* the matrix A:

$$A \overset{\text{def}}{=} \Delta^T \Delta,$$
$$\Delta = [d_{nm}],$$

where d_{nm} is the distance of the nth path through the mth cell.

(4) *Computing* the M-dimensional vector of "true" cell parameters $\mathbf{p}$:

$$\mathbf{p} = [p_m] = [\overline{p_n}\delta_n]\Delta A^{-1},$$

where $\delta_n = \sum_m d_{nm}$ is the total distance for the nth path.

We note that the final accuracy in estimating the true cell parameters $\mathbf{p}$ is dependent on the condition number Λ of A where Λ will be affected by the values of d_{nm}, i.e., the distances of each path through each cell. If we simply increase the number of cells M we find that Λ also increases, resulting in an expected *decrease* in parameter accuracy. Thus, *more* cells without a corresponding increase in the number of SR paths N will produce *degraded* parameter resolution.

A key advantage to this method is that once the "average" parameters have been estimated for each SR path (and by whatever method the user chooses such as the RIGS method, or the simulated annealing method of Fallat and Dosso [FD]), the global range-varying parameter values can be estimated rapidly, simply, and independent of the *number* of parameters computed per path. That is, increasing complexity in the cell environment (via more sediment layers or more complicated bottom sound-speed profiles) does *not* increase the final (Step (2)) global computation. Rather, more parameters will result in an increase in difficulty in the estimation of the *average* path parameters but not in the *final* range-varying parameter estimates.

_____

[1]D. Knobles (personal communication) has suggested that the use of a normal mode code such as ORCA plus a small artificial gradient in the bottom (rather than a false half-space bottom) result in the calculation of leaky modes plus a branch line spectrum as well as the usual propagating modes. This can then produce essentially adiabatic propagation, e.g., for a lossy wedge. He suggests that the usual coupling required for the lossy wedge is an artifact of a false bottom which can be circumvented by means of a slight gradient in the bottom sound-speed profile. See also [WK].

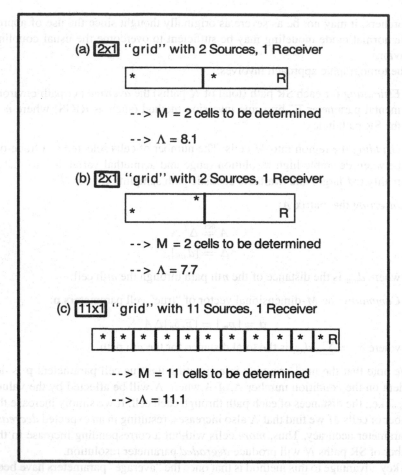

(a) 2x1 "grid" with 2 Sources, 1 Receiver

--> M = 2 cells to be determined

--> Λ = 8.1

(b) 2x1 "grid" with 2 Sources, 1 Receiver

--> M = 2 cells to be determined

--> Λ = 7.7

(c) 11x1 "grid" with 11 Sources, 1 Receiver

--> M = 11 cells to be determined

--> Λ = 11.1

FIGURE 3.2. Ideal two-dimensional scenarios for the application of the tomographic inversion method.

3.3   Results

Our results will concern only Step (2) of the method, i.e., the matrix inversions in range-dependent environments. We shall assume that the "average" per path (slice) parameters have already been determined by some method, e.g., RIGS. The tomography method is insensitive to which parameters are finally estimated (such sensitivity will only affect Step (1) of the method). However, the source-array geometry *will* affect the quality of the tomography inversion (Step (2)).

We begin our calculations with simple SR paths through a 2×1 "grid" placing two sources (*) and one receiver (R) as shown in Figure 3.2(a). We need to estimate an environmental parameter for each of the two cells which make up the two SR paths (one of which is range-dependent). We note that the corresponding inversion condition number Λ is very respectable ($\Lambda = 8.1$) leading to very small errors—if the average parameter values for the two paths are *exact*. In particular, the maximum error (relative to a parameter value of 1.00) will be $1.0 * 10^{-6}$, and the root mean square (rms) error will be $1.0 * 10^{-6}$, both very small numbers as to be expected for the small condition number ($\Lambda \ll 100.0$) of the geometry. While there is no one-to-one mapping between Λ and the rms error, there is an obvious correlation suggesting that small Λ promise small rms errors. If errors are introduced into the average parameter values, then the inversion will degrade correspondingly.

Next we consider a slightly more interesting case where the SR paths are *both* range-variable as shown in Figure 3.2(b). Again we find a very small condition number $\Lambda = 7.7$ leading to maximum and rms errors of $3.0 * 10^{-5}$. We note that the errors are larger than for Figure 3.2(a) even though the condition number is smaller. This indicates that while the condition number is a critical element of the inversion quality, it is not the *only* element. Thus, preestimating inversion quality will not simply be a case of examining Λ.

A final two-dimensional example is seen in Figure 3.2(c) where we now have an 11×1 "grid" with 11 sources, one receiver, and 11 cell parameters to be determined. The corresponding condition number is $\Lambda = 11.1$, and the consequent inversion maximum error for the estimated parameter is $1.0 * 10^{-5}$, the rms error is $4.0 * 10^{-6}$. While this results in the largest Λ of Figure 3.2, the inversion errors are between those of Figures 3.2(a) and 3.2(b).

Of course, the most interesting applications of this method will be for three-dimensional area inversions such as might correspond to the Haro Strait test (we note that details concerning source frequencies, array descriptions, or environmental parameter values are not germaine to the results to follow; see [J] for full information). We have simulated this test scenario with sources and receivers distributed as per the test (see Figure 3.3(a)). We next gridded the area uniformly with 53 cells (see Figure 3.3(b)) and found a Λ of 71,584, a maximum error of $2.5 * 10^{-3}$, and an rms error of $4.0 * 10^{-4}$—assuming perfect accuracy in the estimation of the average parameter values. If the gridding is coarser and set at 14 cells, we arrive at $\Lambda = 8809$, a maximum error of $6.0 * 10^{-4}$, and an rms error of $2.0 * 10^{-4}$ (improved values). If the gridding is set at an even coarser 10 cells, we arrive at $\Lambda = 203$, a maximum error of $2.8 * 10^{-5}$, and an rms error of $1.5 * 10^{-5}$ (even more improved values). We note that it is impossible to declare that $\Lambda = 71,584$ is a "large" value while $\Lambda = 203$ is "small." We can only note that the much larger values of Λ result in larger inversion errors. The appropriateness of the errors depends on *why* one wants the parameter values. Of course, these very small Haro Strait error values are *unrealistic* because:

- we have assumed *perfect* estimates of the average parameter values (see [T1], for a discussion of some error effects); and

- we have *not* accounted for possible mismatch between the "true" environment and the gridded cell structure which may need to be finer or different sizes than assumed. Our simulation examples assumed *perfect* gridding.

44   A. Tolstoy

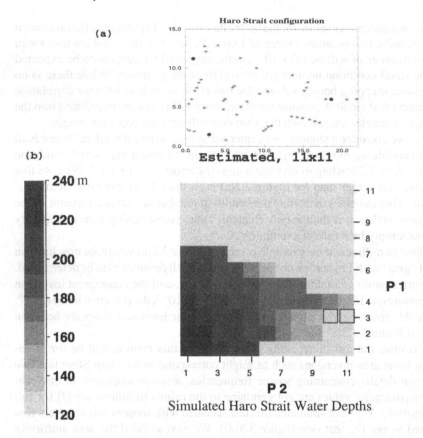

FIGURE 3.3. Simulation of Haro Strait test scenario. (a) Distribution of sources (*) and receiver arrays (•); (b) gridding of region into 53 uniform square cells with water depth indicated by shading (darkest shade is deepest).

3.4   Conclusions

We have begun investigation into a tomographic geoacoustic inversion method appropriate for shallow water. We find that the method is quite general in that in Step (1) *any* user-selected approach can be employed for the estimation of *average* geoacoustic properties per SR path, e.g., RIGS or a simulated annealing approach such as used in the range-independent Workshop97 tests. Then, for Step (2) the method simply involves the inversion a matrix of SR path distances (distances through the *gridded* region of interest) followed by a series of multiplications including a vector of the average SR path parameters. This final computation produces full range and azimuthally varying bottom parameters.

Advantages of the method include:

- The incorporation of a user-selected inversion method for the estimation of average SR path geoacoustic parameters.

- Independence from the number of geoacoustic parameters to be determined per cell. However, this number will be a factor influencing the success of the user-selected inversion method for the SR average parameters.

- Exceptional accuracy for the final estimated parameters—given that the *average* values have been accurately computed by Step (1). That is, Step (2) will introduce only very minor inaccuracies, *if* the geometry of the SR paths well samples the region of interest (as will be suggested by low values of the condition number Λ).

- Application to either two-dimensional range-varying slices or to full three-dimensional volume estimates of environmental parameters.

Disadvantages of the method include:

- The assumption of adiabaticity in the propagation.

- Degradation of results if the region is inappropriately gridded.

- The requirement for multiple sources and multiple receiver/arrays distributed throughout the region of interest.

Acknowledgments. The author would like to thank N.R. Chapman for his efforts on behalf of inversion benchmarking and the Haro Strait data; S. Dosso and A. Caiti for suggestions on matrix regularization (to be pursued in future work); ONR for their continued support for geoacoustic inversion, in general, and this method, in particular; and to M. Taroudakis for his efforts on behalf of inversion resulting in this workshop.

3.5   References

[DYOC] S. Dosso, M.L. Yeremy, J.M. Ozard, and N.R. Chapman. Estimation of ocean-bottom properties by matched field inversion of acoustic field data. *IEEE J. Ocean. Eng.*, **18**:232–239, 1993.

[FD]   M.R. Fallat and S.E. Dosso. Geoacoustic inversion for Worskhop '97 benchmark test cases using simulated annealing. *J. Comput. Acoust.*, **6**(1 & 2):29–44, 1998.

[J]   L. Jaschke. Geophysical Inversion by the Freeze Bath Method with an Application to Geoacoustic Ocean Bottom parameter Estimation. PhD Thesis, University of Victoria, 1997.

[JF]   F.B. Jensen and C.M. Ferla. Numerical solutions of range-dependent benchmark problems in ocean acoustics. *J. Acoust. Soc. Am.*, **87**:1499–1510, 1990.

[JKPS] F.B. Jensen, W.A. Kuperman, M.B. Porter, and H. Schmidt. *Computational Ocean Acoustics*, American Institute of Physics, New York, 1994.

[T1]   A. Tolstoy. Tomographic inversion for geoacoustic parameters in shallow water. *ICTCA'99 Proceedings* book, 2000.

[T2]   A. Tolstoy. MFP benchmark inversions via the RIGS method. *J. Comput. Acoust.*, **6**(1 & 2):185–203, 1998.

[T3]   A. Tolstoy. Performance of acoustic tomography via matched field processing. *J. Comput. Acoust.*, **2**(1):1–10, 1994.

[T4]   A. Tolstoy. *Matched Field Processing in Underwater Acoustics*. World Scientific, Singapore, 1993.

[T5]   A. Tolstoy. Linearization of the matched field processing approach to acoustic tomography. *J. Acoust. Soc. Am.*, **91**(2):781–787, 1992.

[TCB]   A. Tolstoy, N.R. Chapman, and G. Brooke. Workshop97: Benchmarking for geoacoustic inversion in shallow water. *J. Comput. Acoust.*, **6**(1 & 2):1–28, 1998.

[WK]   E.K. Westwood and R.A. Koch. Elimination of branch cuts from the normal mode solution using gradient half-spaces. *J. Acoust. Soc. Am.*, **106**:2513–2523, 1999.

4

Nonlinear Optimization Techniques for Geoacoustic Tomography

Gopu R. Potty
James H. Miller

ABSTRACT Global optimization schemes such as Simulated Annealing (SA) and Genetic Algorithms (GA), which rely on exhaustive searches, have been used increasingly in recent times for the inversion of underwater acoustic signals for bottom properties. Local optimization schemes such as the Levenberg–Marquardt algorithm (LM) and Gauss–Newton methods which rely on gradients, can compliment the global techniques near the global minimum. We use hybrid schemes which combine the GA with LM and Differential Evolution (DE) to invert for the geoacoustic properties of the bottom. The experimental data used for the inversions are SUS charge explosions acquired on a vertical hydrophone array during the Shelf Break Primer Experiment conducted south of New England in the Middle Atlantic Bight in August 1996. These signals were analyzed for their time-frequency behavior using wavelets. The group speed dispersion curves were obtained from the wavelet scalogram of the signals. Hybrid methods mentioned earlier are used for the inversion of compressional wave speeds in the sediment layers. An adiabatic normal mode routine was used to construct the replica fields corresponding to the parameters. Comparison of group speeds for modes 1 to 9 and for a range of frequencies 10 to 200 Hz was used to arrive at the best parameter fit. Error estimates based on the Hessian matrices and a posteriori mean and covariance are also computed. Resolution lengths were also calculated using the covariance matrix. The inverted sediment compressional speed profile compares well with in situ measurements.

4.1 Introduction

Acoustic propagation in shallow water is considerably influenced by the properties of the bottom. Hence methods for accurate and quick estimation of bottom properties have become important especially in shallow waters. Many methods for indirect estimation have been proposed by Tolstoy, Diachok, and Frazer [TDF], Ratilal, Gerstoft, Goh, and Yeo [RGGY], Smith, Rojas, Miller, and Potty [SRMP], Rapids, Nye, and Yamamoto [RNY], and many others. These approaches differ mainly according to the characteristics of the acoustic sources and measurements (travel time, phase, etc.) they utilize for the inversion. When a broadband source is used to generate acoustic energy in a shallow water waveguide, the acoustic propagation exhibits dispersion effects. Group speeds, i.e., the speeds at which energy is transported, differ for different frequencies and modes. In a shallow water waveguide high frequencies generally arrive earlier whereas the low frequencies, which have steeper eigenangles and are important in geoacoustic inversions since they interact with the bottom, arrive later. This dispersion behavior can be utilized for the inversion of bottom properties. Lynch, Rajan, and Frisk [LRF] used a linear perturbation approach to estimate the geoacoustic properties at a location in the Gulf of Mexico based on group speed dispersion characteristics. The perturbation approach breaks down the non-linear problem into a linear one in the vicinity of the solution. Hence, an accurate a priori model of the environment is required to achieve good estimates. This limitation has fueled the increased use of nonlinear methods for inversion, taking advantage of the advanced computational capabilities available at present.

Global optimization schemes such as Simulated Annealing (SA) and Genetic Algorithms (GA) have been used increasingly in recent times for the inversion of underwater acoustic signals for bottom properties. Representative references would be Collins and Kuperman [CK], Hermand and Gerstoft [HG], and Gerstoft [GE]. These methods rely on exhaustive searches, and the time required for the search is often very high. The use of these global approaches does not guarantee that an exact global optimum will be found, even in an infinite amount of time. Local methods can be used to closely examine the region around the best model generated by a global scheme. To achieve this many, investigators have developed hybrid schemes combining global schemes with a local search method. Gerstoft [GE] suggested a combination of global GA and a local Gauss–Newton method. Taroudakis and Markaki [TM] proposed another hybrid scheme wherein the reference environment defined using matched field processing with a GA is subsequently used in connection with a modal phase inversion scheme. This linear modal inversion is meant to fine tune the results obtained through the matched field tomography. We have used two hybrid schemes combining a GA with a Levenberg–Marquardt (LM) method and Differential Evolution (DE) scheme. These two methods are applied within the GA algorithm in order to modify a generation by replacing some of the members of the population.

This chapter is organized as follows. Section 4.2 describes the hybrid inversion schemes used in the present study. The hybrid scheme using the LM method is presented in Section 4.2.1 whereas the DE-based hybrid scheme is presented in Section 4.2.2. These inversion schemes are applied to the data from the Shelf Break Primer Experiment. Section 4.3 presents the details of the Primer Experiment. Section 4.4 describes the analysis of the primer data and the details of the hybrid scheme implementation. Section 4.5 discusses the results including the details of the calculation of resolution lengths. Section 4.6 gives the conclusions of this study.

4.2   Hybrid Schemes

The hybrid schemes presented in this section utilize a GA as their main search engine. GAs are nonlinear optimization schemes, highly efficient in optimizing discontinuous, noisy, highly dimensional, and multimodal objective functions. A simple GA starts with a population of model parameter vectors (individuals) which are randomly generated within the search bounds. These individuals constitute the initial generation in the GA cycle. Successive generations of the individuals are evolved by the application of a set of evolutionary steps (Goldberg, [GO]). Given enough time a GA will usually converge on the optimum, but in practice this is not likely to be a rapid process. Hybrid methods have been developed to overcome this deficiency. In these methods a GA is used to locate the valleys and a more efficient local algorithm climbs down the last few steps to the valley (global minimum). In the hybrid schemes used in our study, we perform minimization using an LM method or a DE scheme to improve the individuals in selected GA generations. Thus some of the individuals in these selected generations get modified by the LM method or the DE scheme before being reinserted into the GA population. These two hybrid implementations are described in detail in the following sections.

4.2.1 Hybrid-LM Scheme

The first hybrid scheme (named hybrid-LM) utilizes the LM method to modify the individuals of the GA population. The LM method is a powerful nonlinear optimization technique which combines the inherent stability of the steepest descent method with the quadratic convergence rate of the Gauss–Newton method. The step direction in the LM method is intermediate between the steepest descent and Gauss–Newton directions. Because of this, the LM method has been proved to be more robust than the Gauss–Newton method. The LM method is used in this inversion scheme in the following manner. A GA inversion is initiated with a predefined population size and number of generations. After a certain number of generations, an LM optimization is performed. The fittest individual in the generation is chosen as the starting vector for this local optimization. This will help to find a local minimum, if any, near the fittest individual. The optimized parameter vector thus obtained is inserted into the GA population replacing the least-fit individual of that generation. Thus the least-fit individual will be replaced with another externally evaluated fitter individual. The frequency of this operation can be varied to obtain better performance with minimum additional computational effort.

4.2.2 Hybrid-DE Scheme

The second hybrid scheme (named hybrid-DE) utilizes the DE scheme to modify the individuals of the GA population. The DE scheme also works with group of model parameter vectors (individuals) like a GA. In the DE method, a new generation is evolved from the previous one by adding the weighted difference between two population members to a third member. If the resulting member yields a lower objective function value than a predetermined population member, the newly generated member replaces the member with which it was compared. The best parameter vector is also evaluated for each generation to keep track of the optimization process. Extracting distance and direction information from the population to generate random deviations results in an adaptive scheme with excellent convergence properties. This hybrid scheme utilizes the DE method to hasten the convergence to the minimum in a GA generation. The DE scheme used in the presented study is described in the following section.

DE Inversion Scheme

There are several variants of DE (Storn and Price, [SP]); the one which is used in the present study is an iteration procedure in which the evolution of the population is controlled by the "distance" between the individuals. Let each generation G consists of NP parameter vectors $x_{i,G}$, $i = 0, 1, 2, 3, \ldots$, NP-1. For each member a trial vector $\mathbf{v}$ is generated according to

$$\mathbf{v} = \mathbf{x}_{r1,G} + F(\mathbf{x}_{r2,G} - \mathbf{x}_{r3,G}), \qquad (4.1)$$

where $r1, r2, r3$ are different integers and are chosen randomly from the interval 0 to NP-1 and $F > 0$. F is a real constant factor which controls the amplification of the differential variation $(\mathbf{x}_{r2,G}-\mathbf{x}_{r3,G})$. Figure 4.1 shows a two-dimensional example that illustrates the different vectors that play a part in the DE scheme. Part of the original population is replaced with $\mathbf{v}$ after comparing with the fitness of the original member. Thus if the individuals of one generation are spread over the whole model space, then the individuals of the next generation will remain the same. But if the individuals are concentrated at a certain point (minimum), then the individuals of the next generation stay at this point until a solution somewhere

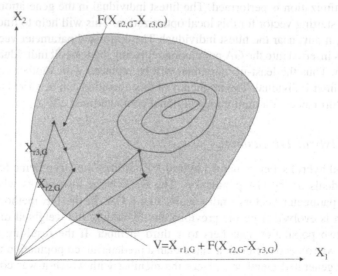

FIGURE 4.1. Two-dimensional example of an objective function showing its contour lines and the process for generating V in the DE scheme.

the case of hybrid-LM. After a certain number of generations, the entire population of GA is used as the starting population for a DE optimization scheme. After a certain number of generations, fitter individuals will be evolved by this adaptive scheme. A certain percentage of the GA population is then replaced with the new individuals evolved by DE. The individuals to be replaced are selected from the least-fit ones. Hence in this method, also least-fit individuals will be replaced with other externally evaluated good individuals.

4.3   Shelf Break PRIMER Experiment

A joint acoustic and physical oceanography experiment, the Shelf Break Primer Experiment, was conducted in the summer of 1996 on the shelf break and continental slope south of New England in the Middle Atlantic Bight. The main objective of the experiment was to study the thermohaline variability and structure of the shelf break front and its effect on acoustic propagation. To accomplish this experimental objective, a combination of acoustic and oceanographic measurements was made. Lynch et al. [LGC] gives the details of these measurements. The acoustic component involved moored tomographic sources and two vertical hydrophone arrays. To study the geoacoustic parameters of the bottom, a number of explosive charges (SUS charges) were also deployed.

The SUS component of the experiment involved the acquisition of broadband acoustic data on two vertical line arrays located on the continental shelf on the northwest vertical line array (NW VLA) and northeast (NE VLA) corners of the experimental area, in water depths of approximately 90 m (Figure 4.2). A P-3 aircraft from the Naval Air Warfare Center in Patuxent Naval Air Station dropped over 80 MK61 explosive charges in an inverted F-pattern as shown in Figure 4.2. These MK61 SUS charges are 0.81 kg of TNT and were set to detonate at a depth of 18 m. Data collected at the NE VLA was used for the present study. The NE VLA consisted of 16 hydrophones which spanned the lower half of the water column. The top hydrophone was at a depth of 45.42 m whereas the bottom one was at 93 m. Data was collected at these hydrophones at a sampling frequency of 1395.09 samples/s.

The sediments in the region of primer experiment consist of an upper layer of Holocene sands, 5–20 m deep, followed by up to 200 m of horizontally stratified layers of silt, sand, and clayey materials from the Pleistocene era (Potty et. al. [PMLS]). While the continental shelf areas surrounding the primer site have been well studied, there is virtually no published geoacoustic data available for the upper 100 m within the actual study region. An Atlantic Margin Coring (AMCOR-6012) Project drill site was located about 20 km due west of the southwestern corner of the experimental area (Figure 4.2). The geotechnical data from this deep core was used to calculate the compressional speed values down to a depth of about 200 m. This AMCOR data was used to compare and validate the inversion results.

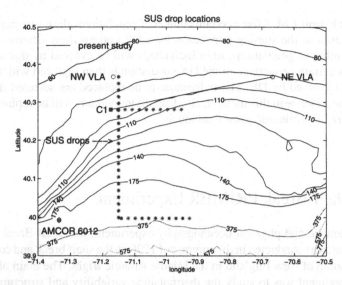

FIGURE 4.2. SUS drop locations at the experimental site. The AMCOR site is shown in the lower left corner of the figure. The propagation path corresponding to the present study is also shown.

4.4   Inversion Using SUS Signals

Acoustic signals collected at NE VLA from SUS charge marked C1 in Figure 4.2 are used for this study. The time series recorded on the middle hydrophone (at a depth of 66.32 m) is shown in Figure 4.3. Dispersion characteristics can be seen even in the raw data. The signals from these explosive sources were then analyzed using wavelets to generate the time-frequency representation of the modal arrivals. The modal dispersion curves obtained using wavelet analysis is shown in Figure 4.4. Modal arrivals corresponding to modes 1 to 9 can be identified in this figure. Early arrivals are very close to one another and difficult to separate. But late arrivals, when they are prominent, are well separated and hence easily to identify. It should be noted that the late arrivals are important to the geoacoustic inversions as they interact with the bottom to a greater extent than early arrivals. It can also be noted that frequency resolution is very poor at higher frequencies (a property of wavelet transform) compared to lower frequencies. This can be improved by adjusting the wavelet parameters which in turn may result in poor time resolution at lower frequencies. Hence different wavelet parameters may be needed to extract the arrival time information at different frequencies. Theoretical group speed curves, calculated based on historical sound speed profiles, were also used to help in identifying the individual mode arrivals. Mode arrival times corresponding to modes 1 to 9 in the frequency range 20 to 200 Hz were identified from the dispersion diagrams, and the corresponding group speed values were used as data for the inversion scheme.

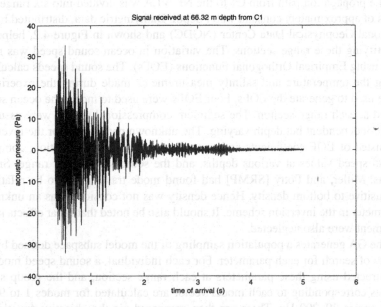

FIGURE 4.3. Time series from SUS C1 recorded at the NE VLA. Location of SUS charge C1 is marked in Figure 4.2.

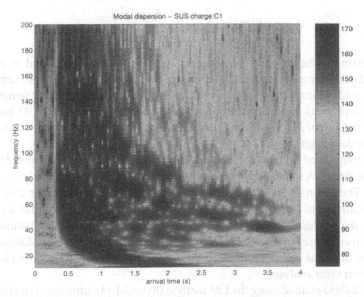

FIGURE 4.4. Wavelet scalogram for the SUS signal shown in Figure 4.3. Modes 1 to 9 can be identified in this figure.

The propagation path from C1 to the NE VLA was divided into six range sections of approximately constant water depth. Bathymetric data, distributed by the National Geophysical Data Center (NGDC) and shown in Figure 4.2, helped in identifying these range sections. The variation in ocean sound speed was modeled using Empirical Orthogonal Functions (EOFs). The sound speeds calculated using the temperature and salinity measurements made during the experiment were used to generate the EOFs. Four EOFs were used to model the ocean sound speed at each range section. The sediment compressional speeds were assumed range-independent but depth varying. The unknown parameters for the inversion consisted of EOF coefficients and water depth at each range section, compressional speed values at various depths, and the source-to-receiver range. Smith, Rojas, Miller, and Potty [SRMP] had found mode travel times to be relatively insensitive to bottom density. Hence density was not considered as an unknown parameter in the inversion scheme. It should also be noted that shear effects in the sediment were also neglected.

The GA generates a population sampling of the model subspace defined by the limits of search for each parameter. For each individual, a sound speed model is constructed using these parameters at each range section, and the group speed values corresponding to each model vector are calculated for modes 1 to 9 and frequencies 20–200 Hz. These are then compared via the objective function to the true group speed values. The objective function for this analysis was based on minimizing the difference between group speed values calculated using the observed time of arrivals and the predicted group speeds

$$E(\mathbf{m}) = \sum_{i=1}^{N} \frac{[d_i - F_i(\mathbf{m})]^2}{\sigma_i^2}, \tag{4.2}$$

where $E(\mathbf{m})$ is the objective function for the mth parameter set, and σ_i is the standard deviation associated with the ith data point. σ_i is calculated approximately as the spread of the spectral peaks in the time-frequency scalograms. The numerator of (4.2) represents the mismatch between the observed data ($\mathbf{d}$, $N \times 1$) and the prediction ($F(\mathbf{m})$, $N \times 1$) of the forward model, where N is the total number of available data points.

The group speed values corresponding to the model vectors are calculated using adiabatic theory. A standard normal mode routine (Porter and Reiss [PR]) was used for this. The group speeds are calculated at each range section and range averaged. The GA inversions were performed using the Genetic Algorithm Matlab toolbox [P] with the stochastic universal sampling selection algorithm, real mutation, and discrete recombination. Parallel GAs were run to make sure that the solution converged to the same minimum. The models sampled by the GA were stored for the a posteriori error analysis.

In the hybrid method using the LM method (hybrid-LM), after the sixteenth and twenty-fourth generations of the GA, the best individual is used to start an LM search. Once LM descends to the local minimum near this individual, the least-fit individual in the GA generation is replaced with this newly obtained individual and the GA operations are continued. By replacing only the bad individual it is hoped that all the good genes in the GA population will be preserved. Thus the GA is allowed to evolve initially without interference until the sixteenth generation. After the two replacements the GA is again run in order to escape from any local minimum. This way the robustness associated with the GA may not be affected much.

The DE, in contrast to the LM, does not depend on gradients to guide its search. It uses scaled differences between existing model vectors to generate new ones. Storn and Price [SP] have claimed excellent convergence properties for this method. In our second hybrid scheme (hybrid-DE), after 20 GA generations the entire GA population is used to initiate a search using DE. DE is allowed to run for 10 cycles. After this the best individual obtained is used to replace the least-fit individual in the GA generation as in the first hybrid scheme. The GA cycle is allowed to continue for another 10 cycles. It should be noted that the GA generation at which the bad individual was replaced (16 and 24 in hybrid-LM and 20 in hybrid-DE) was arrived at after a few trial runs in order to obtain better performance with minimum additional computational effort. It should also be noted that one cycle of the LM operations will introduce forward model computations of the order of 200–300 whereas one cycle of DE will require almost the same number of forward model runs as the GA.

4.5   Results and Discussion

Both hybrid methods seem to perform well compared to the baseline GA method. Figure 4.5 shows the performance of the two hybrid schemes in comparison with the baseline (GA) model. A hybrid-LM method shows a reduction of the objective function when the LM method is applied after generations 16 and 24. But there seems very little further reduction after the twenty-fourth generation by the continued application of the GA. In effect, the hybrid-LM method achieves a fitness level after generation 24, which is equivalent to the fitness at the end of 40 generations of the baseline GA scheme. In the case of hybrid-DE the DE method is applied at generation 20. There is a huge reduction in the objective function due to this. But further application of the GA seems to make no further improvement. Hence, by generation 20, hybrid-DE was able to reach a level which is lower than the level which the baseline model reaches in 40 generations. This saves some computational effort even after considering the additional forward runs associated with DE (which in this case is equivalent to 10 generations of GA runs). It can also be noted that both the hybrid methods perform well compared to the baseline GA scheme but the hybrid-DE performs better than the hybrid-LM method. The resulting compressional wave speed profiles and the standard errors are discussed in the following paragraphs.

Figure 4.6 shows the sediment compressional speeds obtained by the three inversion schemes. In this figure the inversion results for the baseline, and two hybrid

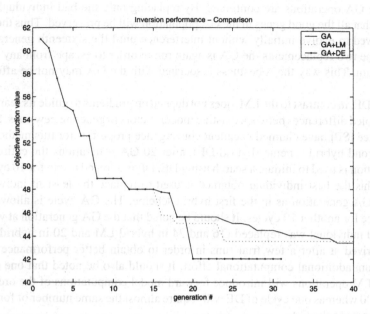

FIGURE 4.5. The comparative performances of the GA, the GA with DE, and the GA with LM.

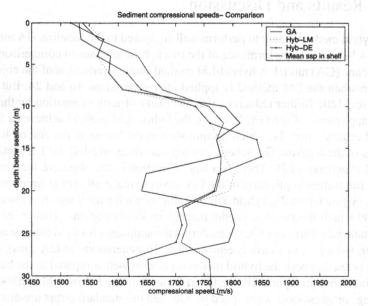

FIGURE 4.6. Compressional wave speed profiles calculated using the three methods. The mean profile is obtained from many inversions using other SUS charge signals.

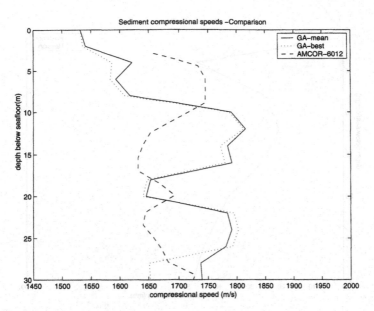

FIGURE 4.7. Compressional wave speed profiles calculated using the GA inversion and using AMCOR-6012 data.

methods, are compared with the mean compressional speed in the shelf region of the experimental area. This mean profile was calculated by performing inversions on different SUS shots distributed over the shelf region (Potty et al. [PMLS]; Potty, Miller, and Lynch [PML]). The results of the GA inversion and the hybrid schemes match well with the mean profile in the shelf, especially in the top 15 m of the sediment. The hybrid-DE appears to be in closer agreement with the mean profile. The GA inversion and the two hybrid methods do not show any sharp differences in trend or in magnitude. The maximum difference is of the order of 25–30 m/s in the top 15 m of the sediment. The average compressional speed in the top 2 m of the sediment is of the order of 1530–1550 m/s which corresponds to silty sediments. Below this depth compressional speeds show an increasing trend down to 15–17 m. This could be attributed to the presence of sandy sediments which are found in the region down to depths of 5–20 m.

Figures 4.7 and 4.8 show the results of the a posteriori analysis done to the model vectors sampled during the baseline GA scheme. The entire population generated by the GA is saved and used to calculate the mean and covariance. Details of this error analysis are discussed by Gerstoft [GE], and Potty et al. [PMLS]. Figure 4.7 shows the sediment compressional speed profile corresponding to the mean and best parameter sets. The mean parameter set is obtained using the a posteriori analysis. The compressional wave speed profile corresponding to the AMCOR-6012 data is also shown in this figure. The AMCOR-6012 data appears to differ considerably from the inversion. It should be noted that the AMCOR data correspond to a location in the slope at a water depth of nearly 300 m (Figure 4.2). It is possible

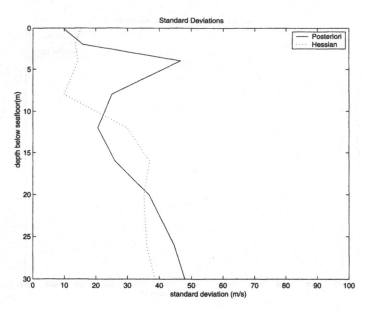

FIGURE 4.8. The standard deviations calculated using a posteriori analysis and using Hessians.

that in the shallower shelf regions where the inversions were done the top layer of sediment is thicker due to increased sedimentation. It can also be noted that the agreement between AMCOR and inversion is better if the AMCOR profile is displaced vertically down by 4 to 5 m. Similar trends were also reported by Potty et al. [PMLS]. The standard deviations are shown in Figure 4.8. Also shown in this figure are the standard deviations calculated using Hessians evaluated locally at the global optimum. Hessians are the second partial derivatives of the objective function with respect to the unknown parameters and are evaluated numerically. This is a measure of curvature of the error surface. When the data are uncorrelated and Gaussian the covariance matrix can be evaluated from the Hessians as shown below (Kopper and Wysession [KW]):

$$[\mathbf{C_m}]_{\text{local}} = \sigma_\mathbf{d}^2 \left[\frac{1}{2} \frac{\partial^2 E}{\partial \mathbf{m}^2} \right]_{\mathbf{m}=\mathbf{m}_{\text{est}}}^{-1}, \tag{4.3}$$

where $\sigma_\mathbf{d}$ is the standard deviation of the error in the data. $\sigma_\mathbf{d}$ is calculated from the width of the spectral peaks in the time-frequency diagram shown in Figure 4.4. The variance of the model error is the diagonal of $\mathbf{C_m}$. Thus the uncertainty in a set of model parameters is the product of the uncertainty in the data and the second-order curvature of the error space about the point $\mathbf{m}_{\text{est}}$.

The standard deviations computed by two different methods agree in magnitude well. Figure 4.9 shows the comparison of the standard deviations of the GA and hybrid methods. The a posteriori error seems too high at 5 m depth for all the methods, indicating high uncertainty around that depth. It is interesting to note

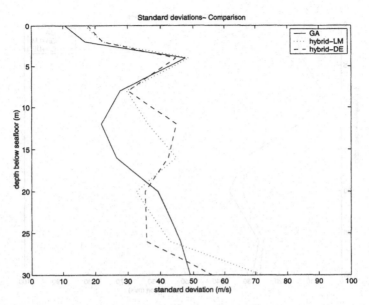

FIGURE 4.9. The standard deviations calculated using a posteriori analysis for the three methods.

that the hybrid methods do not alter the characteristics of the population to an appreciable extent, indicating only minor variations in the standard deviations in the top 10 m. Figure 4.10 shows the standard deviations computed using the Hessians for the three methods. They seem to agree well in trend as well as in magnitude. These standard deviations are calculated based on the Hessians evaluated using the best parameter sets. Their close agreement indicates that the error surfaces corresponding to these methods have identical curvatures near the best parameter values.

4.5.1 Resolution of the Estimates

There is presently no general theory available to evaluate the resolution kernel for nonlinear inverse problems. In practice, one often linearizes the problem around the estimated model and then uses linear inverse theory to make inferences about the resolution and reliability of the estimated models. In one such attempt, Potty and Miller [PM] presented an a posteriori estimation method for resolution by trying to reproduce a small perturbation to the solution. After a satisfactory solution had been obtained, they computed the changes in the data due to a delta-function like perturbation to the model, and inverted the perturbation in the data using the parameters involved in the original solution. The resultant functions are linearized images of the delta-function perturbations and their widths are then indicators of smearing in detail introduced by the inversion process. Potty et al. [PMLS] also used linear perturbation methods to get resolution of the inversion. After obtaining a solution using nonlinear methods a linear perturbation inversion is performed

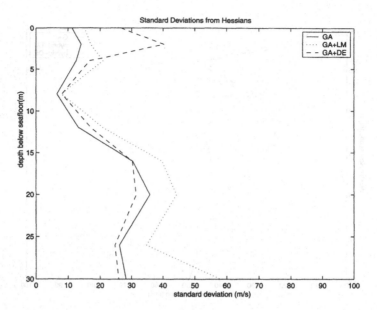

FIGURE 4.10. The standard deviations calculated using Hessians for the three methods.

around this solution. Resolution given by the linear method is taken as an indicator of the reliability of the nonlinear solution.

In this study we used the posterior model covariance to calculate the values of resolution length. A resolution matrix $\mathbf{R}$ can be determined using (Tarantola [T]):

$$\mathbf{R} = \mathbf{I} - \mathbf{C}_{M,\text{prior}}^{-1} \mathbf{C}_M, \qquad (4.4)$$

where $\mathbf{C}_{M,\text{prior}}$ is the prior covariance matrix, whereas $\mathbf{C}_M$ is the posterior model covariance calculated using the models generated using the GA optimization. $\mathbf{I}$ is the identity matrix. The columns of $\mathbf{R}$ give discrete approximation to the resolution kernel, which indicates how the compressional speeds are resolved by the model parameters. It should be noted that the posterior covariance matrix and the resolution kernels are essentially linearized concepts. In cases where different parameter types (compressional speeds, EOF coefficients, water depth, etc.) are used the nondiagonal elements of $\mathbf{R}$ become dimensionally dependent. The resolution matrix for a linear discrete inverse problem is defined as

$$\mathbf{m}_{\text{est}} = \mathbf{R}\mathbf{m}_{\text{true}}, \qquad (4.5)$$

where $\mathbf{m}_{\text{est}}$ and $\mathbf{m}_{\text{true}}$ are the estimated and the true values of the parameters. We can see that the ijth element of $\mathbf{R}$ will have the dimension of parameter i divided by that of parameter j. We can nondimensionalize the ijth element of $\mathbf{R}$ by multiplying it with σ_j/σ_i where σ_i is a representative scale length for parameter i:

$$\mathbf{R}'_{ij} = \mathbf{R}_{ij} \frac{\sigma_j}{\sigma_i}. \qquad (4.6)$$

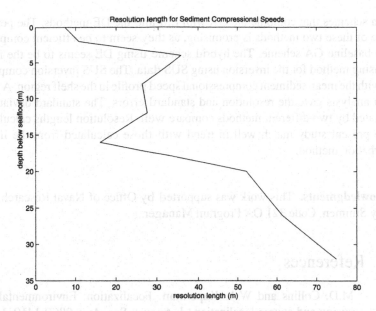

FIGURE 4.11. Model resolution lengths calculated using the a posteriori model covariance matrix.

We have used the roots of the diagonal elements of the prior covariance matrix as the scale length parameters.

The prior covariance matrix can be generated assuming a prior probability density distribution. We have used a constant probability density over the parameter space. Using this assumption we can obtain the elements of the prior covariance matrix as (Sambridge [S]):

$$C_{ij,\text{prior}} = \frac{1}{\sqrt{12}}(u_i - l_i)^2, \quad \text{if} \quad i = j, \qquad (4.7)$$

$$C_{ij,\text{prior}} = 0, \qquad \text{otherwise}, \qquad (4.8)$$

where u_i and l_i are the upper and lower bounds of the ith parameter. Figure 4.11 shows the resolution lengths calculated for the sediment compressional speeds for the top 35 m of the sediments using this approach. The calculated resolution lengths are less than 30 m/s within the top 20 m of the sediment but below this depth they increase to much higher values. The resolution is poor around 5 m depth. These values compare well in trend with the resolution lengths obtained by Potty et al. [PMLS] even though the present method tends to overestimate it.

4.6   Conclusions

Sediment compressional wave speeds were obtained using nonlinear inversion schemes. Data from the Shelf Break Primer Experiment was used in two different hybrid schemes that combine the GA with LM and the DE methods. The performance of these two methods is promising, as they seem to be efficient compared to the baseline GA scheme. The hybrid scheme using DE seems to be the most promising method for the inversion using SUS data. The SUS inversion compares well with the mean sediment compressional speed profile in the shelf region. A posteriori analysis gave the resolution and standard errors. The standard deviations calculated by two different methods compare well. Resolution lengths calculated in the present study match well in trend with those calculated from the linear perturbation method.

Acknowledgments. This work was supported by Office of Naval Research, Dr. Jeffrey Simmen, Code 321 OA Program Manager.

4.7   References

[CK]   M.D. Collins and W.A. Kuperman. Focalization: Environmental focusing and source localization. *J. Acoust. Soc. Am.*, **90**(3):1410–1422, 1991.

[GE]   P. Gerstoft. Inversion of acoustic data using a combination of genetic algorithms and the Gauss–Newton approach. *J. Acoust. Soc. Am.*, **97**(4):2181–2190, 1995.

[GO]   D.E. Goldberg. *Genetic Algorithms in Search, Optimization and Machine Learning.* Addison-Wesley, Boston, MA, 1989.

[HG]   J.P. Hermand and P. Gerstoft. Inversion of broadband multitone acoustic data from the YELLOW SHARK summer experiments. *IEEE J. Ocean. Eng.*, **21**(4):324–346, 1996.

[KW]   K. Koper and M. Wysession. Modelling the Earth's Core and Lowermost Mantle with a Genetic Algorithm. http://levee.wustl.edu/seismology/koper/Papers/JGR96/pkp-ga.html, 1996.

[LGC]   J. Lynch, G. Gawarkiewicz, C. Chiu, R. Pickart, J. Miller, K. Smith, A. Robinson, K. Brink, R. Beardsley, B. Sperry, and G. Potty. Shelfbreak PRIMER—An integrated acoustic and oceanographic field study in the Middle Atlantic Bight. R. Zhang and J. Zhou, (eds.), *Shallow Water Acoust.*, pp. 205–212. China Ocean Press, Beijing, 1997.

[LRF]   J.F. Lynch, S.D. Rajan and G.V. Frisk. A comparison of broad band and narrow band modal inversions for bottom geoacoustic properties at a site near Corpus Christi, Texas. *J. Acoust. Soc. Am.*, **89**(2):648–665, 1991.

[P]   H. Pohlheim. Genetic and Evolutionary Algorithm for use with MATLAB—Version 1.83. http://www.systemtechnik.tu-ilmenau.de/~pohlheim/GA-Toolbox/, 1996.

[PM]   G.R. Potty and J.H. Miller. Geoacoustic tomography: Range dependent inversions on a single slice. *J. Comput. Acoust.*, in press.

[PML]   G.R. Potty, J.H. Miller, and J.F. Lynch. Inversion of sediment geoacoustic properties at the New England Bight. *J. Acoust. Soc. Am.*, in preperation.

[PMLS]   G.R. Potty, J.H. Miller, J.F. Lynch, and K.B. Smith. Tomographic mapping of sediments in shallow water. *J. Acoust. Soc. Am.*, **108**(3):973–986, 2000.

[PR]   M.B. Porter and E.L. Reiss. A numerical method for ocean acoustic normal modes. *J. Acoust. Soc. Am.*, **76**(1):244–252, 1984.

[RGGY]   P. Ratilal, P. Gerstoft, J.T. Goh, and K.P. Yeo. Inversion of pressure data on a vertical array for seafloor geoacoustics properties. *J. Comput. Acoust.*, **6**(122):269–289, 1998.

[RNY]   B. Rapids, T. Nye, and T. Yamamoto. Pilot experiment for the acquisition of marine sediment properties via small scale tomography systems. *J. Acoust. Soc. Am.*, **103**(1):212–224, 1998.

[S]   M. Sambridge. Geophysical inversion with a neighbourhood algorithm-II. Appraising the ensemble. *Geophysics. J. Int.*, **138**:727–746, 1999.

[SP]   R. Storn and A. Price. Differential Evolution—A simple and efficient adaptive scheme for global optimization over continuous spaces. International Computer Science Institute, Technical Report, TR-95-012, 1995.

[SRMP]   K.B. Smith, J.G. Rojas, J.H. Miller, and G. Potty. Geoacoustic inversions in shallow water using direct methods and genetic algorithm techniques. *J. Adv. Mar. Sci. Tech. Soc.*, **4**(2):205–216, 1998.

[T]   A. Tarantola. *Inverse Problem Theory—Methods for Data Fitting and Model Parameter Estimation.* Elsevier, Amsterdam, 1987.

[TM]   M.I. Taroudakis and M.G. Markaki. On the use of matched-field processing and hybrid algorithms for vertical slice tomography. *J. Acoust. Soc. Am.*, **102**(2):885–895, 1997.

[TDF]   A. Tolstoy, O. Diachok, and L.N. Frazer. Acoustic tomography via matched field processing. *J. Acoust. Soc. Am.*, **89**(3):1119–1127, 1991.

5

Estimating the Impulse Response of the Ocean: Correlation Versus Deconvolution

Zoi-Heleni Michalopoulou

ABSTRACT The impulse response of an oceanic waveguide characterizes the propagation effects of the waveguide on a transmitted sequence. When the source signature is known, estimates of the impulse response can be obtained by processing signals received at sensors located in the waveguide, since the received signals are a convolution of the transmitted sequence and the impulse response. This chapter compares two approaches to ocean impulse response estimation, crosscorrelation and deconvolution, in the case of linear and hyperbolic frequency modulated source signals. It is shown that the nature of the transmitted sequence dictates whether correlation or deconvolution is the superior impulse response estimator.

5.1 Introduction

A signal propagating in the ocean is transformed during passage through the underwater medium; the transformation is summarized in a convolution process between the sequence and the ocean impulse response; the convolution process is then distorted by additive noise/interference.

Accurate knowledge of the ocean impulse response is of great value, both to learn more about a specific oceanic site and to implement a computationally efficient variant of space–time coherent detection and estimation, matched-impulse response processing [Miced]. Results presented in [Miced] show matched-impulse response processing to have excellent performance in inversion for source localization, preferable to that of conventional, incoherent matched-field inversion methods. When the source signal is known, obtaining an estimate of the impulse response is intuitively thought of as a matter of signal deconvolution. In some cases, however, the nature of the source signature allows the use of crosscorrelation as a tool for impulse response estimation. In this chapter, we examine the two options, crosscorrelation and deconvolution, for the extraction of the ocean impulse response from data at receiving sensors.

Estimating the ocean impulse response can also be achieved indirectly by first estimating the unknown location and environmental parameters using matched field processing [Tol93] or a model-based matched-filter (mbmf) [HR88], [Her98]. Once estimates of the unknown parameters are obtained, they can be used to form an estimate of the ocean impulse response using a broadband sound propagation model. The indirect approach is not discussed here, but details on matched field processing and model-based matched-filtering can be found in [Tol93, HR88, Her98].

The structure of the chapter is as follows. Section 5.2 describes the source sequences of interest (linear frequency modulated (lfm) and hyperbolic frequency modulated (hfm) signals). Sections 5.3 and 5.4 discuss deconvolution and cross-correlation in impulse response estimation for lfm and hfm source signals. Section 5.5 compares the two estimation methods, whereas Section 5.6 presents impulse response estimates from the SWellEX-96 data. Section 5.7 summarizes and concludes the presented work.

5.2   Source Signals

The source signals employed in this chapter are tailored to the specifications of the SWellEX-96 experiment [SWe96]. During part of the experiment, a source transmitted lfm and hfm sequences. The frequency content of the signals was between 200 and 400 Hz, the transmitted signal duration was 5 s, and the sampling rate was 1500 Hz. Spectrograms of the transmitted sequences are shown in Figure 5.1. The SWellEX-96 oceanic environment was characterized by impulse responses with a duration of about 0.2 s. This duration requires 301 samples for the impulse response representation for a 1500 Hz sampling rate, whereas the 5 s duration of the transmitted signals requires 7501 samples. The received time series, being the convolution of source and impulse response, is represented with 7801 samples and

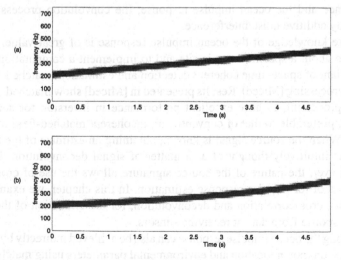

FIGURE 5.1. Spectrograms of the (a) lfm and (b) hfm transmitted signals.

is slightly longer than the 5 s transmitted signal because of the spreading effect of the ocean impulse response on the signal.

5.3   Signal Deconvolution

A sequence, $r(t)$, measured at a receiving phone is related to the transmitted signal, $s(t)$, by the convolution

$$r(t) = \int_{-\infty}^{\infty} s(\tau)h(t - \tau)\,d\tau + n(t), \tag{5.1}$$

where $h(t)$ is the impulse response of the ocean, and $n(t)$ is the noise, here assumed to be white and Gaussian. Sequence $h(t)$ depends on the source and receiver locations and the physics of the environment (water column depth, seafloor sediment properties, sound speed profile, etc.). In active sonar problems the input sequence $s(t)$ is known, and interest is on estimating the impulse response $h(t)$ or parameters related to it. Since the received sequence is the result of the convolution of $h(t)$ and $s(t)$ (plus the noise distortion), the problem of extracting $\hat{h}(t)$, an estimate of $h(t)$, from the received signal is a problem of signal deconvolution.

Deconvolution is readily accomplished using a Wiener filter [Hay96]: Wiener deconvolution for ocean impulse response extraction solves a simple linear system, relating the autocorrelation matrix R_{ss} of the transmitted sequence and the unknown impulse response of the ocean to the crosscorrelation between the transmitted and received signals. Mathematical formulation of the Wiener deconvolution reduces to a system of normal equations:

$$R_{ss}\hat{h} = y, \tag{5.2}$$

where $\hat{h}$ corresponds to the estimate of the impulse response $h(t)$ obtained from this system and y is the crosscorrelation sequence between $s(t)$ and $r(t)$.

When the autocorrelation matrix R_{ss} is nonsingular, the system (5.2) can be solved by direct matrix inversion as follows:

$$\hat{h} = R_{ss}^{-1}y. \tag{5.3}$$

In our problem, however, R_{ss} is nearly singular. The signal carries information with a frequency content between 200 and 400 Hz, whereas the sampling rate is 1500 Hz. Frequency components below 200 Hz and above 400 Hz are represented by very small eigenvalues, leading to singularity problems. Thus, direct inversion produces numerically unstable results. Such ill-conditioning is typical in deconvolution problems; a better solution is required.

Singular value decomposition may be used to yield numerically stable solutions to ill-conditioned problems such as deconvolution [RS90]. Decomposing the nearly singular coefficient matrix R_{ss} of the system described in (5.2), we can make the deconvolution results more robust to noise, by eliminating small singular values that cause the instability problem in the solution of the linear system.

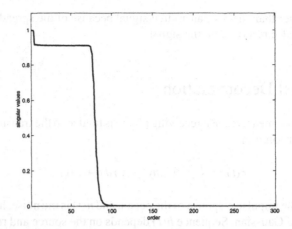

FIGURE 5.2. Singular values obtained from applying singular value decomposition to the autocorrelation matrix of an lfm sequence.

Applying singular value decomposition, the symmetric, positive definite matrix R_{ss} can be written as

$$R_{ss} = U \Sigma V^T, \tag{5.4}$$

where Σ is a diagonal matrix carrying singular values, that is, the eigenvalues of matrix R_{ss}. Because R_{ss} is a positive definite matrix, $U = V$, where matrix U carries the eigenvectors of R_{ss}; in a qualitative sense, U carries elemental waveforms that are weighted by the singular values of matrix Σ to form R_{ss}.

Then (5.3) becomes

$$\hat{\mathbf{h}} = V \Sigma^{-1} U^T \mathbf{y}, \tag{5.5}$$

leading to equations of the form $\hat{h}_i = V(U^T \mathbf{y})_i / \sigma_i$, for the ith element of vector $\hat{\mathbf{h}}$.

Figure 5.2 shows the singular values (normalized with respect to the largest) from the decomposition of matrix R_{ss}, the autocorrelation matrix of an lfm sequence. The singular values decrease significantly after, approximately, the first 90 values. Similar observations can be made from Figure 5.3, showing the singular values from decomposition of the autocorrelation matrix of an hfm sequence. In this case, singular values become negligible at around order 80.

Figures 5.4 and 5.5 show the Fourier transforms of the columns of matrix V resulting from singular value decomposition of the autocorrelation matrices for the lfm and hfm signals, respectively. It can be observed that the first 90 columns for the lfm signal and the first 80 for the hfm signal have frequency content between 200 and 400 Hz, the frequency range of the transmitted signals. After these columns, however, the frequency content expands to frequencies below 200 and above 400 Hz. The thresholds after which frequency content is outside the desired range coincide with the thresholds after which singular values become negligible. As discussed in [RS90], one can then ignore all columns of matrix V that have frequency content outside the range of interest, which is equivalent to ignoring the

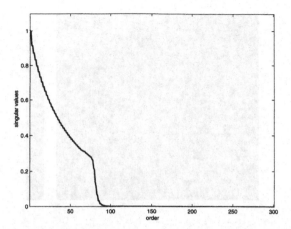

FIGURE 5.3. Singular values obtained from applying singular value decomposition to the autocorrelation matrix of an hfm sequence.

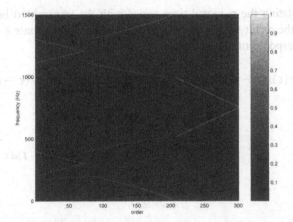

FIGURE 5.4. Fourier transform of the columns of matrix V resulting from singular value decomposition applied to the autocorrelation matrix of an lfm sequence.

effects of the very small singular values in the denominator of the linear system solution of (5.4). This action is effective in high SNR situations. When the noise level is high, more columns of matrix V should be discarded. In this way we discard small singular values that are comparable to or smaller than the noise level.

5.4   Crosscorrelation in Deconvolution

The autocorrelation of several broadband source sequences approaches an impulse. We can exploit this impulse-like autocorrelation to estimate the ocean impulse response without resorting to the complex process of deconvolution.

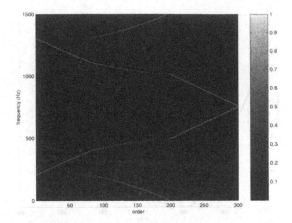

FIGURE 5.5. Fourier transform of the columns of matrix V resulting from singular value decomposition applied to the autocorrelation matrix of an hfm sequence.

Crosscorrelating the received time series with the transmitted broadband sequence $s(t)$, the result will be the convolution of an approximate δ function and the impulse response of the medium

$$
\int_{-\infty}^{\infty} r(\tau)s(\tau - t)\,d\tau = \int_{-\infty}^{\infty} \left[\int_{-\infty}^{\infty} s(\tau - t')h(t')\,dt' \right] s(\tau - t)\,d\tau
$$
$$
+ \int_{-\infty}^{\infty} n(\tau)s(\tau - t)\,d\tau
$$
$$
= \int_{-\infty}^{\infty} h(t') \left[\int_{-\infty}^{\infty} s(\tau - t')s(\tau - t)\,d\tau \right] dt'
$$
$$
+ \int_{-\infty}^{\infty} n(\tau)s(\tau - t)\,d\tau. \tag{5.6}
$$

Integral $\int_{-\infty}^{\infty} s(\tau - t')s(\tau - t)\,d\tau$ is the autocorrelation of the source signal. For a high Signal-to-Noise Ratio (SNR) and when this autocorrelation approximates an impulse

$$
\hat{h}(t) = \int_{-\infty}^{\infty} r(\tau)s(\tau - t)\,d\tau \approx \int_{-\infty}^{\infty} h(t')\delta(t - t')\,dt' = h(t), \tag{5.7}
$$

where $\hat{h}(t)$ is the estimated ocean impulse response.

Naturally, this method will result in a good estimate of the impulse response if the match between the autocorrelation of $s(t)$ and an impulse is very good. The quality of the match, therefore, depends on the nature of the source sequence. Below we investigate the autocorrelation properties of lfm and hfm signals.

For a broadband sequence one can calculate the effective bandwidth β_e [Bur84]:

$$
\beta_e = \frac{R^2(0)}{\int_{-\infty}^{\infty} |R(\tau)|^2\,d\tau}, \tag{5.8}
$$

where $R(\tau) = \int_{-\infty}^{\infty} s(t)s(t - \tau)\,dt$, the autocorrelation of signal $s(t)$. The value of the effective bandwidth characterizes the similarity between the autocorrelation function of a sequence and an impulse. In our problem, we compare the effective bandwidth to the actual bandwidth of the transmitted source sequences; similarity between effective and actual bandwidth shows an impulse-like autocorrelation behavior, whereas disparity between the two bandwidth values indicates deviation of the autocorrelation from an impulse.

Analytical calculation of the effective bandwidth is not straightforward for lfm and hfm sequences. It can, however, be numerically approximated and it was found to be 200 and 172 Hz, respectively, for the sequences studied here. The effective bandwidth value of 200 demonstrates the very close proximity of the lfm autocorrelation to an impulse. The autocorrelation of the hfm pulse is less similar to a delta function; the effective bandwidth has a value smaller than 200, which is the bandwidth of the hfm pulse. thus, the crosscorrelation method is expected to work better for impulse response estimation for the lfm rather than the hfm sequence considered here.

These observations are supported by Figure 5.6, which shows the average crosscorrelation between estimated and true impulse responses in a synthetic, shallow-water environment versus SNR; the impulse response estimation is performed using the crosscorrelation process described above applied to lfm and hfm source sequences. It is readily seen that the performance of the crosscorrelator in impulse response estimation is superior for the lfm sequences. The performance gap decreases as the SNR obtains lower values since, under these circumstances, the noise swamps the deviation of the autocorrelation function of the sequence from an impulse.

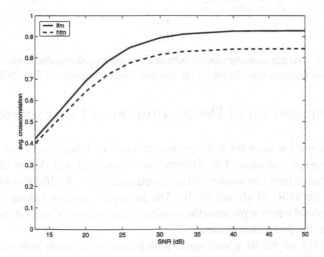

FIGURE 5.6. Average crosscorrelation between estimated impulse responses obtained through the crosscorrelation process and true impulse responses for lfm and hfm source sequences.

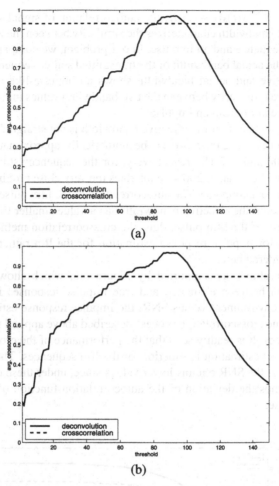

(a)

(b)

FIGURE 5.7. Average crosscorrelation between the estimated impulse response through deconvolution and crosscorrelation for (a) lfm and (b) hfm sequences. The SNR is 40 dB.

5.5 Comparison of Deconvolution and Crosscorrelation

Figures 5.7 and 5.8 show the average crosscorelation obtained between estimated and true impulse responses. The estimates were obtained with deconvolution and crosscorrelation between source and received time series for lfm and hfm source sequences for SNRs of 40 and 20 dB. The average crosscorrelations are plotted versus threshold which represents the number of columns of V used in the solution to the linear deconvolution system.

For an SNR of 40 dB a very good match can be achieved between the true and estimated impulse response. The average crosscorrelation between the two reaches 0.98 for the SVD deconvolution process and 0.93 for the crosscorrelation process for the lfm case; the corresponding values are 0.97 and 0.84 for the hfm

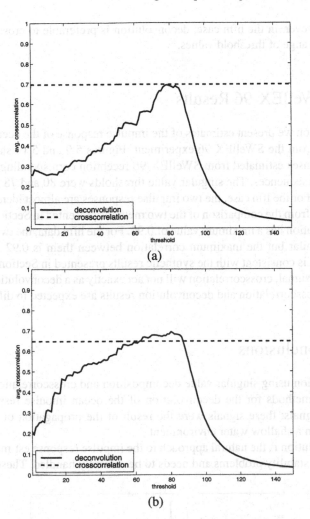

FIGURE 5.8. Average correlation between the estimated impulse response through deconvolution and crosscorrelation for (a) lfm and (b) hfm sequences. The SNR is 20 dB.

case. The reason that the two methods are very close when the source sequence is linear frequency modulated, is that, as discussed in Section 5.4, the autocorrelation of the lfm pulse is very similar to an impulse. Thus, the crosscorrelation works very well as an impulse response estimator. Matters are different in the hfm case. There is greater disparity between the autocorrelation of the hfm pulse and an impulse and, consequently, the crosscorrelation process yields inferior estimates of the impulse response. These observations are reinforced by Figure 5.8. For the lower SNR (which is still quite high), the deconvolution and crosscorrelation results are comparable only for one particular threshold value; this would make the crosscorrelation preferable to the deconvolution process, because a subtle change in the threshold choice would affect significantly (and negatively) the deconvolution result. However, in the hfm case, deconvolution is preferable to crosscorrelation for a wide range of threshold values.

5.6   SWellEX-96 Results

In this section we present estimates of the impulse response of the ocean obtained from data from the SWellEX-96 experiment. Figures 5.9 and 5.10 show the impulse responses estimated from SWellEX-96 reception corresponding to lfm and hfm source sequences. The singular value thresholds were 80 and 78 for the two sequences. For the lfm case, the two impulse responses are almost identical which is expected from the comparison of the two methods presented in Section 5.5; their crosscorrelation has a maximum value of 0.99. For the hfm data, the two estimates are still similar but the maximum correlation between them is 0.92. Again this observation is consistent with the synthetic results presented in Section 5.5; for an hfm source signal, crosscorrelation will not act exactly as a deconvolution process, and thus crosscorrelation and deconvolution results are expected to differ.

5.7   Conclusions

Deconvolution using singular value decomposition and crosscorrelation were examined as methods for the deconvolution of the ocean impulse response from recorded signals; these signals were the result of the propagation of broadband sequences in a shallow water environment.

Deconvolution is the natural approach to the impulse response estimation problem but has stability problems and needs to be carefully handled. These problems

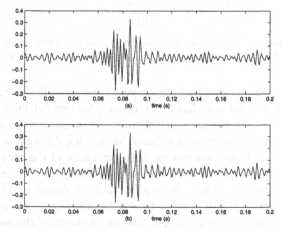

FIGURE 5.9. Estimated impulse response, with (a) crosscorrelation and (b) deconvolution for an lfm source sequence.

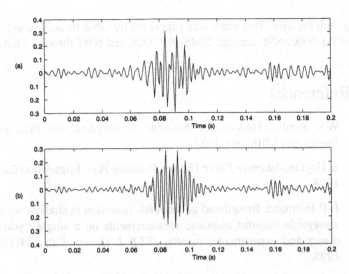

FIGURE 5.10. Estimated impulse response, with (a) crosscorrelation and (b) deconvolution for an hfm source sequence.

can be addressed by applying singular value decomposition to the autocorrelation matrix of the source sequence and rejecting small singular values. Use of singular value decomposition, however, implies a criterion that allows us to keep or reject singular values. Deconvolution results are, typically, very sensitive to this criterion.

In order to avoid thresholding problems in deconvolution with singular value decomposition, crosscorrelation is here offered as an alternative for impulse response estimation. Crosscorrelation yields good impulse response estimates when the autocorrelation of the source sequence is close to an impulse. Our work showed that lfm sequences such as those employed in the SWellEX-96 experiment have autocorrelation functions adequately close to an impulse for crosscorrelation to work very effectively as a deconvolution method. Crosscorrelation does not work as well with hfm sequences because of the deviation of their autocorrelation from an impulse. In the hfm case, the impulse response estimates obtained with crosscorrelation are inferior to regular deconvolution.

Using crosscorrelation when the source sequence does not have an impulse-like autocorrelation or employing deconvolution with singular value decomposition using a suboptimum threshold leads to disparities between the estimated and true impulse responses. Future work will address the effect of these disparities in inversion techniques that make use of impulse response estimates. Specifically, we will attempt to quantify the performance loss that can be tolerated in deconvolution for accurate localization and geoacoustic inversion to be feasible.

Acknowledgments. The author wishes to thank Dr. Newell Booth for providing the SWellEX-96 data, Dr. W. Hodgkiss and Dr. James Murray for their invaluable help with the data. This work was supported by ONR through grant number N00014-97-1-0600, NSF through DMS-9872008, and NJIT through SBR.

5.8 References

[Bur84]   W.S. Burdic. *Underwater Acoustic System Analysis*. Prentice Hall, Englewood Cliffs, NJ, 1984.

[Hay96]   S. Haykin. *Adaptive Filter Theory*. Prentice-Hall, Englewood Cliffs, NJ, 1996.

[Her98]   J.-P. Hermand. Broadband geoacoustic inversion in shallow water from waveguide impulse response measurements on a single hydrophone; theory and experimental results. *IEEE J. Ocean. Eng.*, 24(1):41–66, 1998.

[HR88]   J.-P. Hermand and W.I. Roderick. Delay-Doppler resolution performance of large time-bandwidth-product linear FM signals in a multipath ocean environment. *IEEE J. Ocean. Eng.*, 94(5):1709–1727, 1988.

[Miced]   Z.-H. Michalopoulou. Matched impulse response processing for shallow water localization and geoacoustic inversion. *J. Acoust. Soc. Am.*, submitted.

[RS90]   E.J. Rothwell and W. Sun. Time domain deconvolution of transient radar data. *IEEE Trans. Antennas Propagation*, 38(4):470–475, 1990.

[SWe96]   SWellEX-96 Preliminary Data Report. Technical report, 1996.

[Tol93]   A. Tolstoy. *Matched Field Processing for Underwater Acoustics*. World Scientific, Singapore, 1993.

6

Regularized Inversion for Towed-Array Shape Estimation

Stan E. Dosso
Nicole E. Collison

ABSTRACT This chapter describes a new approach to the inverse problem of estimating the shape of a ship-towed hydrophone array using near-field acoustic measurements. The data consist of the relative travel times of arrivals along direct and reflected paths from sources deployed by two consort ships maintaining station with the moving tow ship (the "dual-shot method"). Previous inversion algorithms typically apply least-squares methods based on simplifying assumptions, such as straight-line propagation and exact knowledge of the source positions. Here, a regularized inversion is developed based on ray theory, with the source positions included as unknown parameters subject to a priori estimates and uncertainties. In addition, a minimum-structure array shape is determined by minimizing the three-dimensional curvature subject to fitting the data to a statistically meaningful level, thereby reducing spurious fluctuations (roughness) in the solution. Finally, the effect of the survey geometry is investigated by defining a mean sensor-position error measure based on the a posteriori uncertainty of the inversion. The optimal source configuration is determined by minimizing this error with respect to the source positions using an efficient hybrid optimization algorithm. The inversion and optimization procedures are illustrated using realistic synthetic examples.

6.1 Introduction

Towed arrays consist of a line of hydrophones housed in a neutrally buoyant, acoustically transparent hose, and are commonly used in military sonar systems and marine seismic exploration. For optimal array-processing performance, it is important to determine the relative positions of the sensors (or, equivalently, the shape of the array), a problem commonly referred to as array element localization (AEL). For example, a general rule to achieve a loss of less than 1 dB in array-processing gain requires the sensor positions be known to within $\lambda/10$, where λ is the wavelength at the frequency of interest [H+96]. The shape of a towed array is dynamic, and is influenced by changes in the tow ship's course and speed and by shear currents in the water column. Array-mounted instruments, such as depth and heading sensors, can provide information on array shape. Typically, such instruments are distributed at intervals along the array, and do not provide measurements of the positions of individual sensors. Sensor locations can be obtained by smoothly interpolating between measured positions (e.g., employing a cubic spline), or by applying hydrodynamic modeling which attempts to account for the dynamic behavior of an array with known properties (weight, drag, etc.) under towing conditions.

Van Ballegooijen et al. [vB⁺89] introduced an acoustic approach to towed-array shape estimation which does provides individual sensor positions. This method can be used to independently verify array shapes derived from depth and heading measurements, or can be applied in conjunction with these measurements (it is not, however, a practical approach for all operational scenarios). The approach, known as the "dual-shot method," makes use of explosive sources deployed in the water column by two consort ships maintaining station with the moving tow ship. The suggested survey configuration is to have the consort ships 500–1000 m away from the array and spaced at an angle of 90° relative to the array center. The measured data ideally consist of relative travel times of the acoustic arrivals along direct, bottom-reflected, and bottom–surface reflected paths; however, in practical cases, one or other of the reflected arrivals may not be usable. Assuming the source positions and ocean sound speed are known, the travel-time data can be inverted to infer the positions of the array sensors. A reference hydrophone installed on the tow ship serves to locate the array with respect to the ship. The source depths should be approximately equal to the array depth, and the survey is ideally carried out at a deep-water site with a flat bottom. This ensures that the direct arrivals travel horizontally while the reflected arrivals travel nearly vertically. This survey geometry is designed to provide good three-dimensional (3-D) sensor localization, with the direct-path arrivals providing horizontal (x–y) control, and the reflected arrivals providing vertical (z) control.

The acoustic inversion applied to the dual-shot method [vB⁺89] represents an application of the least-squares method (minimizing the squared data error). Other approaches to similar AEL inversion problems are described in [H⁺96], [CD82], [Mil83], [SH90], [B⁺96], and [OC97]. This chapter develops a new approach for towed-array shape estimation. The algorithms presented here represent an extension of the approach to AEL recently developed for moored horizontal and vertical hydrophone arrays in [D⁺98b], [D⁺98a], and [DS99]. A number of sources of error, neglected in most previous AEL inversions, are addressed in the algorithms developed here. For instance, AEL inversion algorithms typically treat the source positions as known parameters. However, the inevitable errors in source positions are often nonnegligible, and in some cases cause larger inversion errors than the uncertainties of the travel-time data (i.e., source-position errors can represent the limiting factor in AEL inversion). Therefore, the source locations are not treated as known quantities here, but rather are included as unknown parameters (subject to a priori estimates with uncertainties) in the inversion algorithm. Another source of error often neglected involves the curvature of acoustic ray paths due to the depth-dependence of the ocean sound-speed profile. To address this, a general raytracing-based inversion algorithm is developed here. Third, errors in the measured sound-speed profile affect the accuracy of AEL inversion. Sound-speed measurements are generally accurate in a relative sense, but can suffer from bias errors of up to 2 m/s due to inaccurate calibration [VH98]. Hence, the sound-speed bias is also included as a (constrained) unknown in the inversion. It should be noted that it is also possible for the ocean sound speed to vary both laterally and temporally in an unknown manner during an AEL survey; however, these uncertainties are not constrained by the acoustic data and will not be considered here.

The inversion algorithm is formulated to include both the travel-time data and available a priori information. In addition to prior estimates for some parameters (source positions, sound-speed bias), the a priori information also includes the physical expectation that the array shape is essentially smooth (i.e., does not involve small-scale roughness or fluctuations). This is applied by minimizing the three-dimensional curvature (roughness) of the array, subject to fitting the data to a statistically meaningful level, to obtain a minimum-structure solution. This essentially applies a priori information about the correlation between sensor positions, rather than information about the positions themselves. Minimizing the array curvature, subject to fitting the data, is physically reasonable, since the effect of towing the array (even while turning) and the stiffness of the array-housing tube typically preclude excessive roughness (e.g., sharp zig-zags in array shapes). In addition, this procedure produces the simplest array shape that is consistent with the data. Any deviations from a straight array are definitely required by the data, and are not artifacts of the inversion algorithm. Seeking minimum-structure solutions is philosophically consistent with Occam's Razor, and in geophysical inversion is often referred to as Occam's inversion [C+87]. In contrast, the least-squares method typically over-fits the data, in effect fitting the noise as well as the data, which can lead to unphysically rough solutions that complicate acoustic signal analysis.

A final issue concerns the use of relative travel-time measurements in AEL inversion. Two approaches are possible here. The first approach is to remove the source instants from the problem by considering appropriate differences between the relative travel times as the data to be inverted. The alternative is to treat the relative travel times as the data, and include each source instant as an unknown parameter to be determined in the inversion. The latter approach is adopted here since it results in data with smaller uncertainties, and since it allows the inversion algorithm greater scope in the application of a priori information.

AEL represents a nonlinear inverse problem that is inherently nonunique, and a closed-form solution does not exist. An effective approach is based on local linearization and iteration, reducing the nonlinear problem to a series of linear inversions that can be solved using methods of linear inverse theory. Section 6.2.1 considers linearization of the AEL inversion, and Secction 6.2.2 describes linearized inversion with application of a priori information via the method of regularization. In addition to determining the source positions, an estimate of the error in these positions is derived. These error estimates can be applied to the problem of designing optimal AEL surveys by minimizing the sensor-position error with respect to the source positions, as described in Section 6.2.3. The ray-theory basis for the inversion and optimization algorithms is outlined in Section 6.2.4.

In Section 6.3, the array-shape inversion and optimal survey design are illustrated with a series of synthetic examples.

6.2 Theory

6.2.1 Linearization

The set of N acoustic arrival times t measured in an AEL survey can be written in general vector form as

$$t = T(m) + n. \tag{6.1}$$

In (6.1), the forward mapping T represents the arrival times of the acoustic signals along ray paths between sources and receivers (an explicit expression for T and an efficient method of computing ray arrival times is given in Section 6.2.4). The model m of M unknown parameters consists of three-dimensional position variables x_i, y_i, z_i for each sensor, position variables x_j, y_j, z_j and source instant t_j^0 for each source, and the sound-speed bias c_b. Finally, n represents the data errors (noise). Equation (6.1) defines the AEL inverse problem: given a dataset of measured arrival times t and knowledge of the forward mapping T, determine the model parameters m which gave rise to the data. As mentioned previously, this inverse problem is nonlinear; however, a local linearization is obtained by expanding $T(m) = T(m_0 + \delta m)$ in a Taylor series to first order about an arbitrary starting model m_0 to yield

$$t = T(m_0) + J\delta m, \tag{6.2}$$

where δm represents an unknown model perturbation and J is the Jacobian matrix consisting of the partial derivatives of the data functionals with respect to the model parameters

$$J_{kl} = \partial T_k(m_0)/\partial m_l \tag{6.3}$$

(partial derivatives of the ray travel time are derived in Section 6.2.4). Defining $\delta t = t - T(m_0)$, the expansion can be written

$$J\delta m = \delta t. \tag{6.4}$$

Equation (6.4) defines a linear inverse problem for δm which can be solved using methods of linear inverse theory. Once δm is determined, the corresponding model solution is $m = m_0 + \delta m$. Since nonlinear terms are neglected in (6.4), the model m may not adequately reproduce the measured data. In this case, the starting model is updated, $m_0 \leftarrow m$, and the inversion is repeated iteratively until an acceptable solution is obtained.

Least-squares methods are typically applied to invert (6.4), provided the inversion is well-posed. For ill-conditioned inversions, some form of minimum-norm solution is usually applied (i.e., the perturbation $|\delta m|^2$ is minimized at each iteration), such as singular value decomposition [CD82] or Levenberg–Marquardt

[Mil83], [OC97] inversion. This approach has been referred to as the "creeping" method [S+90] since it iteratively progresses toward a solution by a series of small perturbations, with the final model retaining a dependence on the initial starting model. Note that since the linear inverse problem (6.4) is formulated in terms of the model perturbation (not the model), a priori information about the model cannot be included directly in the inversion.

An alternative to the creeping method can be formulated by substituting $\delta\mathbf{m} = \mathbf{m} - \mathbf{m}_0$ into expansion (6.4) to obtain [Old83]:

$$\mathbf{Jm} = \delta\mathbf{t} + \mathbf{Jm}_0 \equiv \mathbf{d}. \tag{6.5}$$

This expression relates known quantities (the right side, which may be considered modified data $\mathbf{d}$) directly to $\mathbf{m}$: the linearized inverse problem is formulated in terms of the model, not the model perturbation. In this case, a priori information regarding the model can be applied directly to the inversion, often leading to a more physically meaningful solution [Old83]. This approach has been referred to as the "jumping" method [S+90], since the size of the model change at each iteration is not minimized and the final solution is generally independent of the starting model. The jumping method was first applied to AEL inversion in [D+98a] and [D+98b]; this approach is also followed here.

6.2.2 Regularized Inversion

To consider the linear inverse problem (6.5), assume that the error n_i on datum t_i is due to an independent, Gaussian distributed random process with zero mean and standard deviation ν_i. The least-squares solution for a system of linear equations (6.5) is found by minimizing the χ^2 misfit

$$\chi^2 = |\mathbf{G}(\mathbf{Jm} - \mathbf{d})|^2 \tag{6.6}$$

with respect to the model $\mathbf{m}$, where $\mathbf{G} = \mathrm{diag}[1/\nu_1, \ldots, 1/\nu_N]$ weights the data according to their uncertainties. The solution, determined by setting $\partial\chi^2/\partial\mathbf{m} = 0$, is

$$\mathbf{m} = \left[\mathbf{J}^T\mathbf{G}^T\mathbf{G}\mathbf{J}\right]^{-1}\mathbf{J}^T\mathbf{G}^T\mathbf{G}\mathbf{d}. \tag{6.7}$$

The least-squares approach provides an unbiased solution, provided the matrix to be inverted is nonsingular. In addition to the possibility of singularity, the matrix can be ill-conditioned, leading to an unstable inversion (small errors on the data lead to large errors on the solution). Ill-posed (singular or ill-conditioned) inverse problems result when the data do not contain enough linearly independent information to fully constrain the solution. In AEL, the source–receiver geometry essentially determines the conditioning of the inverse problem. Including the source positions, as well as the sensor positions, as unknowns always leads to an ill-posed inversion.

The method of regularization provides a particularly useful approach to ill-posed linear inversions. Regularization is based on formulating a unique, stable inversion by explicitly including a priori information regarding the solution. This is accomplished by minimizing an objective function ϕ which combines a term representing the χ^2 data misfit and a regularizing term that imposes the a priori expectation that the model $\mathbf{m}$ in some manner resembles a prior estimate $\hat{\mathbf{m}}$:

$$\phi = |\mathbf{G}(\mathbf{Jm} - \mathbf{d})|^2 + \mu |\mathbf{H}(\mathbf{m} - \hat{\mathbf{m}})|^2. \tag{6.8}$$

In (6.8), $\mathbf{H}$ is a weighting matrix known as the regularization matrix (described below), and the variable μ serves as a trade-off parameter controlling the relative importance assigned to the data misfit and the a priori expectation in the minimization. The regularized solution is obtained by minimizing ϕ with respect to $\mathbf{m}$ to yield

$$\mathbf{m} = \hat{\mathbf{m}} + \left[\mathbf{J}^T \mathbf{G}^T \mathbf{GJ} + \mu \mathbf{H}^T \mathbf{H}\right]^{-1} \left[\mathbf{J}^T \mathbf{G}^T \mathbf{Gd} - \mathbf{J}\hat{\mathbf{m}}\right]. \tag{6.9}$$

The presence of the term $\mu \mathbf{H}^T \mathbf{H}$ within the square brackets in (6.9) ensures that the matrix to be inverted is well conditioned.

The regularization matrix $\mathbf{H}$ in (6.8) and (6.9) provides considerable flexibility in the application of a priori information in the inversion. For instance, if prior model parameter estimates $\hat{\mathbf{m}}$ are available, an appropriate weighting is given by [vS89]:

$$\mathbf{H} = \text{diag}[1/\xi_1, \ldots, 1/\xi_M], \tag{6.10}$$

where ξ_j represents the uncertainty for the jth parameter estimate $\hat{m}_j$. This weighting correctly applies prior parameter estimates which can vary over orders of magnitude in uncertainty. An alternative form of regularization is to apply a priori information to derivatives of the model parameters [C+87], [P+92]. For instance, if the a priori expectation is that the parameters are well approximated by a smooth function, then an appropriate choice is $\hat{\mathbf{m}} = \mathbf{0}$ and

$$\mathbf{H} = \begin{bmatrix} -1 & 2 & -1 & 0 & 0 & 0 & 0 & \cdots & 0 \\ 0 & -1 & 2 & -1 & 0 & 0 & 0 & \cdots & 0 \\ \vdots & & & & \ddots & & & & \vdots \\ 0 & \cdots & 0 & 0 & 0 & -1 & 2 & -1 & 0 \\ 0 & \cdots & 0 & 0 & 0 & 0 & -1 & 2 & -1 \end{bmatrix}. \tag{6.11}$$

For this choice of $\hat{\mathbf{m}}$ and $\mathbf{H}$, $\mathbf{H}(\mathbf{m} - \hat{\mathbf{m}})$ represents a discrete approximation to the second derivative of $\mathbf{m}$, and the regularization term in (6.8) provides a measure of the total curvature or roughness R of the model:

$$R = |\mathbf{H}(\mathbf{m} - \hat{\mathbf{m}})|^2. \tag{6.12}$$

Applying this regularization minimizes the model roughness, producing a minimum-structure solution. In each case, the regularization is appropriately applied by choosing the trade-off parameter μ so that the χ^2 data misfit achieves its expected value of $\langle \chi^2 \rangle = N$ for N data [P+92], thereby applying the a priori information subject to ensuring that the data are fit to a statistically appropriate level.

The AEL inverse problem considered here involves both types of a priori information described above. In particular, prior parameter estimates for the source locations are available from the consort ship navigation, and the prior estimate for the sound-speed bias is zero. The expectation that the towed array shape is smooth can be applied by minimizing the three-dimensional curvature (roughness). To simultaneously apply two different types of a priori information to a linear inverse problem, an augmented objective function can be formed which includes two regularization terms [D+98a]:

$$\phi = |\mathbf{G}(\mathbf{Jm} - \mathbf{d})|^2 + \mu_1|\mathbf{H}_1(\mathbf{m} - \hat{\mathbf{m}}_1)|^2 + \mu_2|\mathbf{H}_2(\mathbf{m} - \hat{\mathbf{m}}_2)|^2. \quad (6.13)$$

In (6.13), the first regularization term is taken to represent the a priori parameter estimates for the source locations and sound-speed bias. Hence, $\hat{\mathbf{m}}_1$ consists of the prior estimates for these parameters, with zeros for the remaining parameters. The regularization matrix $\mathbf{H}_1$ is of the form of (6.10) with diagonal elements consisting of the reciprocal of the estimate uncertainty for parameters with prior estimates, and zeros for the remaining parameters. The second regularization term is taken to represent the a priori expectation of a smooth array shape. Hence, $\hat{\mathbf{m}}_2$ is taken to be zero, and $\mathbf{H}_2$ is of the form of (6.11) for the sensor position parameters, with rows of zeros corresponding to the remaining parameters. Rows of zeros are also included in $\mathbf{H}_2$ at appropriate locations to separate the measures of curvature in x, y, and z. In this case, minimizing (6.13) leads to the solution

$$\mathbf{m} = \hat{\mathbf{m}}_1 + \left[\mathbf{J}^T\mathbf{G}^T\mathbf{GJ} + \mu_1\mathbf{H}_1^T\mathbf{H}_1 + \mu_2\mathbf{H}_2^T\mathbf{H}_2\right]^{-1}\left[\mathbf{J}^T\mathbf{G}^T\mathbf{Gd} - \mathbf{J}\hat{\mathbf{m}}_1\right]. \quad (6.14)$$

The AEL inversion algorithm consists of an iterative application of (6.14), initiated from an arbitrary starting model. Convergence of the algorithm is based on two criteria: (i) obtaining a misfit to the measured data of $\chi^2 = N$ for N data; and (ii) obtaining a stable solution such that the rms (root-mean-square) change in the sensor positions between iterations is $\Delta < 0.1$ m. Regarding the first criterion, note that although (6.14) is derived based on the χ^2 misfit for the linear inverse problem (6.5) that approximates the nonlinear problem (6.1) at each iteration, the convergence of the inversion algorithm must be judged in terms of the misfit to the nonlinear problem

$$\chi^2 = |\mathbf{G}(\mathbf{T}(\mathbf{m}) - \mathbf{t})|^2. \quad (6.15)$$

An equivalent, and sometimes more convenient, measure is the rms data misfit

$$X = \left[\chi^2/N\right]^{1/2}, \quad (6.16)$$

with an expected value $\langle X \rangle = 1$.

The most subtle aspect of implementing the inversion has to do with assigning values to the two trade-off parameters, μ_1 and μ_2, which control the balance between the data misfit and the two types of a priori information. An effective procedure [D+98a] is to set

$$\mu_2 = \alpha\mu_1 \quad (6.17)$$

for a fixed value of α, and determine the value of μ_1 at each iteration which yields the desired χ^2 misfit (discussed below). The final model obtained from this procedure can then be examined to ascertain whether the value of α was appropriate based on a comparison of the parameter residuals. This comparison can be quantified by defining the rms misfit associated with the a priori estimates

$$\hat{X} = [|\mathbf{H}_1(\mathbf{m} - \hat{\mathbf{m}}_1)|^2 / \hat{M}]^{1/2}, \tag{6.18}$$

where $\hat{M}$ is the number of model parameters with a priori estimates. To fit the prior estimates within their uncertainties, $\hat{X} \approx 1$. If $\hat{X} \ll 1$, then a smaller value of α is required; if $\hat{X} \gg 1$, a larger value of α is required. The inversion can be repeated with a new value of α until $\hat{X} \lesssim 1$ is achieved. In practice, determining an appropriate value for α is a straightforward procedure, typically requiring only two or three trial inversions. The value for α need not be refined too highly, since the uncertainties of the parameter estimates are only approximate, and the sensor positions recovered in the inversion are not generally sensitive to changes in α of less than about a factor of 2.

The above procedure reduces the problem of determining two trade-off parameters to a one-dimensional search for the parameter μ_1 that produces the desired rms misfit X at each iteration. The trade-off parameter μ_1 is chosen so that X is reduced by approximately a factor of 100 at each iteration until $X = 1$ is achieved. Controlling the change in misfit in this manner limits the change in the model at each iteration. This helps ensure that the linear approximation is valid, and stabilizes the convergence. Since X increases monotonically with μ_1, it is straightforward to determine the value for μ_1 which produces the desired X at a given iteration. At early iterations an approximate value for μ_1 is sufficient, and a bisection algorithm is employed. Near convergence, the bisection solution is improved by applying one or more iterations of Newton's method to determine a precise value for μ_1.

Finally, for a linear inverse problem and Gaussian noise, the marginal a posterior probability distribution for the ith model parameter is also Gaussian, with mean equal to the inversion result and variance given by the ith diagonal entry of the solution covariance matrix

$$\mathbf{C} = \langle (\mathbf{m} - \langle \mathbf{m} \rangle)(\mathbf{m} - \langle \mathbf{m} \rangle)^T \rangle. \tag{6.19}$$

Substituting from (6.14) into (6.19) leads (after some manipulation) to

$$\mathbf{C} = [\mathbf{J}^T \mathbf{G}^T \mathbf{G} \mathbf{J} + \mu_1 \mathbf{H}_1^T \mathbf{H}_1 + \mu_2 \mathbf{H}_2^T \mathbf{H}_2]^{-1}. \tag{6.20}$$

This expression includes the effect of the smoothness regularization (μ_2 term). However, it can be argued that this regularization represents a somewhat arbitrary prior assumption which is difficult to quantify, and that it is preferable to omit this term and define the covariance matrix

$$\mathbf{C} = [\mathbf{J}^T \mathbf{G}^T \mathbf{G} \mathbf{J} + \mu_1 \mathbf{H}_1^T \mathbf{H}_1]^{-1}. \tag{6.21}$$

In practice, it is generally found that (6.20) and (6.21) produce very similar values, although (6.21) always yields slightly larger variances (i.e., is a more conservative estimate). Finally, the expected standard deviation σ_i of parameter m_i is given by

$$\sigma_i = \sqrt{C_{ii}}. \tag{6.22}$$

The parameter error estimates σ_i depend on the data uncertainties through $\mathbf{G}$, on the uncertainties in the a priori parameter estimates through $\mathbf{H}_1$, and on the source–receiver geometry through $\mathbf{J}$. The above uncertainty analysis applies to linear inverse problems; however, for linearized inversions such as AEL, it can provide meaningful estimates of the expected uncertainty [DS99]. A particularly useful application of the a posterior uncertainty is in optimal survey design, considered in the following section.

6.2.3   Optimal Experiment Design

The previous section developed an inversion algorithm for localizing the sensors of a towed array using sources deployed by two escort ships. This section considers the related problem of determining the optimal survey geometry, i.e., the configuration of source positions that produces the most accurate inversion for sensor positions. To this end, an AEL error measure is defined based on the a posterior uncertainties of the recovered sensor positions from (6.21) and (6.22). The optimal survey configuration can then be determined by minimizing this error measure with respect to the source positions [DS99]. Note that this procedure is based on an *expected* inverse problem (i.e., the array configuration, data errors, and parameter uncertainties are assumed, but there are no data to invert). In this case, an appropriate value of the trade-off parameter μ_1 in (6.21) can be determined by carrying out a number of representative synthetic inversions with randomly generated errors and uncertainties.

A number of different AEL error measures can be defined using the standard deviations of the individual sensor-position parameters given by (6.22). Let σ_x, σ_y, σ_z represent the standard deviations of the x, y, z Cartesian coordinates of the sensor positions and let $\sigma_r = [\sigma_x^2 + \sigma_y^2 + \sigma_z^2]^{1/2}$. The error measure that is considered here is

$$E = \frac{1}{N_S} \sum_r \sigma_r, \tag{6.23}$$

where N_S is the total number of sensors to be localized. This measure represents the mean three-dimensional error of the sensor positions. The source configuration that minimizes this error measure will provide the sensor-position estimates that are the most accurate on average. Alternatively, the maximum sensor-position error can be minimized [DS99]. Other error measures can also be devised and may be appropriate for specific AEL objectives. For example, if accurate sensor depths are deemed more important than accurate horizontal positions, the σ_z term in the definition of σ_r could be weighted by a factor greater than one and E minimized; however, such cases are not considered further here.

Optimal AEL survey design consists of determining the set of source-position parameters that minimizes the sensor-position error E. This is a strongly nonlinear minimization problem which typically has a large number of local minima, and is not amenable to linear optimization methods. Global optimization methods, such as simulated annealing (SA) [HS94] and genetic algorithms (GA) [HB98], have been applied to minimization problems associated with geophysical experiment design, but can be relatively inefficient. Recently, hybrid optimization methods have been developed and applied to geophysical [L$^+$95] and ocean-acoustic [FD99] inverse problems. Hybrid methods combine local and global approaches to produce a more efficient optimization. Here, a hybrid optimization algorithm that combines the local downhill simplex (DHS) method with SA is applied to optimal AEL survey design. For completeness, the following subsections briefly describe SA, DHS and the hybrid simplex simulated annealing (SSA) algorithms. For more details, see [DS99].

Simulated Annealing (SA)

SA is a global optimization method that can be applied to minimize a function E with respect to a set of model parameters defined on a given search interval. The algorithm consists of a series of iterations involving random perturbations of the parameters. After each iteration a control parameter, the temperature T, is decreased slightly. Perturbations that decrease E are always accepted; perturbations that increase E are accepted conditionally, with a probability P that decreases with T according to the Boltzmann distribution

$$P(\Delta E) = \exp(-\Delta E / T). \qquad (6.24)$$

Accepting some perturbations that increase E allows the algorithm to escape from local minima in search of a better solution. As T decreases, however, accepting increases in E becomes increasingly improbable, and the algorithm eventually converges. The rate of reducing T and the number and type of perturbations define the annealing schedule. The method of fast SA (FSA) [SH76] is based on using a Cauchy distribution to generate the parameter perturbations and reducing the width of the distribution with the temperature. The narrow peak and flat tails of the Cauchy distribution provide concentrated local sampling together with occasional large perturbations, allowing a faster rate of temperature reduction than standard SA.

Downhill Simplex (DHS)

Global optimization methods widely search the parameter space and avoid becoming trapped in unfavorable local minima. However, since individual steps are computed randomly, these methods can be quite inefficient at moving downhill. In contrast, local (gradient-based) methods move efficiently downhill, but typically become trapped in a local minimum close to the starting model. The DHS method [P$^+$92] is a local inversion technique based on a geometric scheme for moving downhill in E that does not require the computation of partial derivatives or the solution of systems of equations. DHS navigates the search space using a simplex of $M + 1$ models in an M-dimensional parameter space (e.g., Figure 6.1(a), for

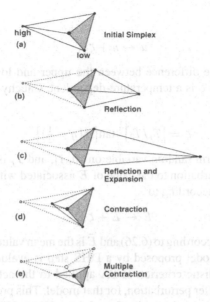

(a)
low
Initial Simplex

(b)
Reflection

(c)
Reflection and
Expansion

(d)
Contraction

(e)
Multiple
Contraction

FIGURE 6.1. DHS steps in three dimensions (after [P+92]).

$M = 3$). The algorithm initially attempts to improve the model with the highest value of E by reflecting it through the opposite face of the simplex (Figure 6.1(b)). If the new model has the lowest E in the simplex, an extension by a factor of 2 in the same direction is attempted (Figure 6.1(c)). If the model obtained by the reflection still has the highest E, the reflection is rejected and a contraction by a factor of 2 along this direction is attempted (Figure 6.1(d)). If none of these steps decrease E, then a multiple contraction about the lowest-E model is performed (Figure 6.1(e)). This process is repeated until the value of E for each model of the simplex converges to a common value (i.e., the simplex shrinks to a single point).

Simplex Simulated Annealing (SSA)

The goal of hybrid inversion is to combine local and global methods to exploit the advantages of each (i.e., to move efficiently downhill, yet avoid becoming trapped in local minima). Here, a hybrid SSA inversion is described that incorporates the local DHS method into a global SA search. Unlike standard SA, the SSA inversion operates on a simplex of models rather than on a single model, and instead of employing purely random model perturbations, DHS steps with a random component are applied to perturb the models. To introduce the random component, the DHS steps are not computed directly from the current simplex of models, but rather from a secondary simplex which is formed by applying random perturbations to all the model parameters and E values associated with the current simplex. The perturbations to the current simplex used to produce the secondary simplex are computed using a Cauchy distribution and reducing the distribution width with temperature as follows. Each source-position parameter $u \in \{x, y, z\}$ is perturbed according to

$$u \leftarrow u + \zeta \delta, \tag{6.25}$$

where δ represents the difference between the upper and lower limits assumed for u and the quantity ζ is a temperature-dependent, Cauchy-distributed random variable computed as

$$\zeta = [T_j/T_0]^{\frac{1}{2}} \tan[\pi(\eta - \frac{1}{2})]. \tag{6.26}$$

In (6.26), η is a uniform random variable on [0, 1], and T_j is the temperature at the jth step. The perturbation to the value of E associated with each model in the simplex is computed according to

$$E \leftarrow E + \zeta \bar{E}, \tag{6.27}$$

where ζ is computed according to (6.26) and $\bar{E}$ is the mean value of E for the current simplex. Each new model proposed by a DHS step is evaluated for acceptance based on the probabilistic criterion of SA applied to the actual (not perturbed) energies, before and after perturbation, for that model. This provides a mechanism for accepting uphill steps and escaping from local minima. If any DHS step results in parameter values outside their given search interval, the parameters are set to the interval bound prior to evaluation. After the set of perturbations is complete, the temperature is reduced according to

$$T_j = \beta^j T_0, \tag{6.28}$$

where β is a constant less than one. An appropriate starting temperature T_0 can be determined by requiring that at least 90% of all perturbations are accepted initially. Appropriate values for β and the number of perturbations per temperature step are usually straightforward to determine with some experimentation.

At high temperatures where the random component of the perturbations dominates, the SSA method resembles an FSA global search. At low temperatures, where the random component is small, the method resembles the local DHS method. At intermediate temperatures, the method makes a smooth transition between these two endpoints. The efficiency of the algorithm can be improved further by "quenching" the optimization when it approaches convergence (i.e., when E effectively stops decreasing) by switching to a pure DHS algorithm to avoid the slow final convergence typical of SA. A block diagram illustrating the basic SSA algorithm is given in Figure 6.2.

6.2.4   Ray Travel Times and Derivatives

For completeness, this subsection describes the classical ray theory applied to compute the acoustic travel times and partial derivatives required in the inversion and optimal survey design algorithms of the previous subsections. Consider an acoustic source and receiver in the ocean at (x_j, y_j, z_j) and (x_i, y_i, z_i), respectively, with $z_j < z_i$ (source above receiver is assumed in the equations given here; for

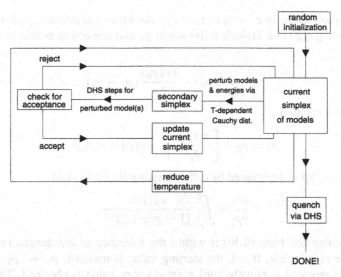

FIGURE 6.2. Block diagram illustrating the SSA algorithm (after [FD99]).

the reverse, a negative sign is required in all integrals unless otherwise noted). The horizontal range between source and receiver is given by

$$r = \sqrt{(x_i - x_j)^2 + (y_i - y_j)^2}. \qquad (6.29)$$

Expressions for the range r and arrival time T along a (nonturning) ray path between source and receiver are easily derived by applying Snell's law to an infinite stack of infinitesimal layers [T+76]:

$$r = \int_{z_j}^{z_i} \frac{pc(z)\,dz}{\left[1 - p^2 c^2(z)\right]^{1/2}}, \qquad (6.30)$$

$$T = t^0 + \int_{z_j}^{z_i} \frac{dz}{c(z)\left[1 - p^2 c^2(z)\right]^{1/2}}, \qquad (6.31)$$

where t^0 represents the source instant. In (6.30) and (6.31), the ray parameter $p = \cos \theta(z)/c(z)$ is constant along a ray path, and defines the take-off (grazing) angle at the source. The ray parameter for an eigenray connecting source and receiver is usually determined by searching for the value of p which produces the correct range (to a specified tolerance) using (6.30). An efficient procedure of determining p for direct-path eigenrays is based on Newton's method [D+98a]. An initial estimate p_0 is calculated assuming straight-line propagation with a sound speed c_H representing the harmonic mean of the measured sound-speed profile between source and receiver

$$c_H = (z_i - z_j) \left/ \int_{z_j}^{z_i} \frac{dz}{c(z)} \right. \qquad (6.32)$$

(this equation holds for $z_j < z_i$ or $z_i < z_j$). An improved estimate p_1 is obtained by expanding $r(p)$ in a Taylor's series about p_0 and neglecting nonlinear terms to give

$$r(p) = r(p_0) + \frac{\partial r(p_0)}{\partial p}(p_1 - p_0), \tag{6.33}$$

which has a solution

$$p_1 = p_0 + \left[\frac{\partial r(p_0)}{\partial p}\right]^{-1} (r(p) - r(p_0)). \tag{6.34}$$

In (6.34), $\partial r / \partial p$ is determined by differentiating (6.30) to yield

$$\frac{\partial r}{\partial p} = \int_{z_j}^{z_i} \frac{c(z)\,dz}{[1 - p^2 c^2(z)]^{3/2}}. \tag{6.35}$$

If $r(p_1)$ computed from (6.30) is within the tolerance of the desired range, the procedure is complete. If not, the starting value is updated, $p_0 \leftarrow p_1$, and the procedure repeated iteratively until a satisfactory value is obtained. The travel time along the ray path is then computed using (6.31). Since Newton's method converges quadratically near the solution, this is an efficient method of determining direct eigenrays to high precision.

In addition to computing travel times, the linearized inversion algorithm requires partial derivatives of travel time with respect to source and receiver coordinates, source instant, and sound-speed bias. Consider first the partial derivative with respect to horizontal coordinate x_i. Employing the chain rule

$$\frac{\partial T}{\partial x_i} = \frac{\partial T}{\partial p}\frac{\partial p}{\partial r}\frac{\partial r}{\partial x_i} = \frac{\partial T}{\partial p}\left[\frac{\partial r}{\partial p}\right]^{-1}\frac{\partial r}{\partial x_i}. \tag{6.36}$$

The three partials on the right side of (6.36) can be calculated from (6.31), (6.30), and (6.29), respectively, yielding

$$\frac{\partial T}{\partial x_i} = p(x_i - x_j)/r. \tag{6.37}$$

Similarly, partial derivatives with respect to the other horizontal coordinates are

$$\frac{\partial T}{\partial x_j} = p(x_j - x_i)/r, \tag{6.38}$$

$$\frac{\partial T}{\partial y_i} = p(y_i - y_j)/r, \tag{6.39}$$

$$\frac{\partial T}{\partial y_j} = p(y_j - y_i)/r. \tag{6.40}$$

The partial derivative of T with respect to vertical coordinate z_i can be determined by differentiating (6.31) to give

$$\frac{\partial T}{\partial z_i} = \int_{z_j}^{z_i} \frac{pc(z)\,dz}{[1 - p^2 c^2(z)]^{3/2}}\left(\frac{\partial p}{\partial z_i}\right) - \frac{1}{c(z_i)\left[1 - p^2 c^2(z_i)\right]^{1/2}}. \tag{6.41}$$

An expression for $\partial p / \partial z_i$ can be obtained by noting that

$$\frac{\partial r}{\partial z_i} = 0 = \int_{z_j}^{z_i} \frac{c(z)\,dz}{[1 - p^2 c^2(z)]^{3/2}} \left(\frac{\partial p}{\partial z_i}\right) - \frac{pc(z_i)}{[1 - p^2 c^2(z_i)]^{1/2}}. \tag{6.42}$$

Solving for $\partial p / \partial z_i$ and substituting into (6.41) yields

$$\frac{\partial T}{\partial z_i} = \frac{1}{c(z_i)} [1 - p^2 c^2(z_i)]^{1/2}. \tag{6.43}$$

Similarly,

$$\frac{\partial T}{\partial z_j} = -\frac{1}{c(z_j)} [1 - p^2 c^2(z_j)]^{1/2}. \tag{6.44}$$

To account for bias in the measured sound-speed profile, let $c(z) = c_t(z) + c_b$, where $c_t(z)$ is the true sound speed and c_b is the bias. Differentiating (6.31) with respect to c_b (and noting $\partial p / \partial c = -p/c$) leads to

$$\frac{\partial T}{\partial c_b} = -\int_{z_j}^{z_i} \frac{dz}{c^2(z)\,[1 - p^2 c^2(z)]^{1/2}}. \tag{6.45}$$

Finally, the derivative of T with respect to the source instant t^0 in (6.31) is simply given by

$$\frac{\partial T}{\partial t^0} = 1. \tag{6.46}$$

To implement numerically the equations derived above, it is assumed that a digital sound-speed profile can be represented by a series of layers with a linear sound-speed gradient in each layer. The simplest ray paths to trace involve bottom and/or surface reflections, since these do not involve turning points (i.e., points where the ray passes through zero grazing angle and changes vertical direction as the result of refraction). Sea surface and bottom reflections are modeled using the method of images, i.e., representing the reflected path by adirect ray path from an image source located above the surface or below the bottom, respectively. Ray paths involving both surface and bottom reflections require both an image source and an image receiver. To apply the method of images, the sound-speed profile is reflected about the interfaces in the same manner as the sources. In the following, let $\{(z_k, c_k)\}$ represent the sound-speed profile including the requisite reflections for a particular reflected path, let $\{c_k'\}$ be the corresponding sound speed gradients, and let z_j and z_i be the source and receiver depths, respectively. For the case of linear sound-speed gradients, the integrals in equations (6.30), (6.31), (6.35), and (6.45) can be evaluated analytically, yielding the following results, where $w_k \equiv (1 - p^2 c_k^2)^{1/2}$:

$$r = \sum_{k=j}^{i-1} \frac{w_k - w_{k+1}}{pc_k'}, \tag{6.47}$$

$$T = t^0 + \sum_{k=j}^{i-1} \frac{1}{c'_k} \left[\log_e \frac{c_{k+1}(1 + w_k)}{c_k(1 + w_{k+1})} \right], \tag{6.48}$$

$$\frac{\partial r}{\partial p} = \sum_{k=j}^{i-1} \frac{w_k - w_{k+1}}{p^2 c'_k w_k w_{k+1}}, \tag{6.49}$$

$$\frac{\partial T}{\partial c_b} = \sum_{k=j}^{i-1} \frac{1}{c'_k} \left[\frac{w_{k+1}}{c_{k+1}} - \frac{w_k}{c_k} \right]. \tag{6.50}$$

Calculating integrals along direct ray paths is somewhat more complicated, since these paths can involve turning rays. Rather than simply integrating (summing) along the ray path from source to receiver as per the reflected rays above, the possibility that the direct ray turns must be checked as it enters each layer along its path. Consider the case of a downward-propagating ray entering the kth layer. The turning depth for this ray is given by

$$z_T = z_k + (1/p - c_k)/c'_k. \tag{6.51}$$

If this depth is less than z_{k+1} (the bottom of the kth layer) the ray turns; if not, it proceeds into layer $k + 1$. If the direct ray does not turn between source and receiver, (6.47)–(6.50) apply. However, if the direct ray turns in layer t, then the correct procedure involves four steps:

(i) integrate from the source depth z_j down to z_t (the top of the tth layer) using the above equations;

(ii) integrate from z_t to the turning depth z_T (where, by definition, $w_T = 0$);

(iii) integrate upward from z_T to z_t; and

(iv) integrate upward from z_t to the receiver depth z_i. Applying this procedure leads to the following equations for turning rays:

$$r = \sum_{k=j}^{t-1} \frac{w_k - w_{k+1}}{pc'_k} + \frac{2w_t}{pc'_t} + \sum_{k=t}^{i+1} \frac{w_k - w_{k-1}}{pc'_{k-1}}, \tag{6.52}$$

$$T = t^0 + \sum_{k=j}^{t-1} \frac{1}{c'_k} \left[\log_e \frac{c_{k+1}(1 + w_k)}{c_k(1 + w_{k+1})} \right]$$

$$+ \frac{2}{c'_t} \log_e \frac{1 + w_t}{pc_t} + \sum_{k=t}^{i+1} \frac{1}{c'_{k-1}} \left[\log_e \frac{c_{k-1}(1 + w_k)}{c_k(1 + w_{k-1})} \right], \tag{6.53}$$

$$\frac{\partial r}{\partial p} = \sum_{k=j}^{t-1} \frac{w_k - w_{k+1}}{p^2 c'_k w_k w_{k+1}} - \frac{2}{c'_t p^2 w_t} + \sum_{k=t}^{i+1} \frac{w_k - w_{k-1}}{p^2 c'_{k-1} w_k w_{k-1}}, \tag{6.54}$$

$$\frac{\partial T}{\partial c_b} = \sum_{k=j}^{t-1} \frac{1}{c'_k} \left[\frac{w_{k+1}}{c_{k+1}} - \frac{w_k}{c_k} \right] + \frac{2w_t}{c_t c'_t} + \sum_{k=t}^{i+1} \frac{1}{c'_{k-1}} \left[\frac{w_{k-1}}{c_{k-1}} - \frac{w_k}{c_k} \right]. \tag{6.55}$$

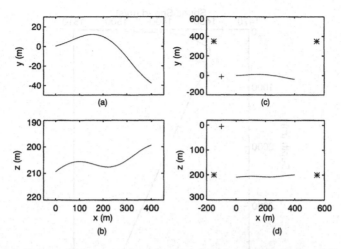

FIGURE 6.3. Source-receiver geometry for the synthetic examples. (a) and (b) show the towed array in $x–y$ and $x–z$ planes, respectively. (c) and (d) show the same at a larger scale, and include the positions of the sources (asterisks) and the ship-mounted reference hydrophone (cross).

6.3 Examples

6.3.1 Inversion for Towed-Array Shape

This subsection illustrates the regularized inversion algorithm for towed-array shape estimation with a number of synthetic examples. Figure 6.3 shows the source–receiver geometry for the examples. The array consists of 41 sensors, each nominally separated by 10 m, for a total array length of approximately 400 m. The array is curved in the horizontal ($x–y$) plane, as shown in Figure 6.3(a), representing the effect of a course change by the tow ship (assumed to be to the left of the array, see Figure 6.3(c)). The total horizontal deflection of the array is approximately 40 m. The array tilts generally upward (from fore to aft) in the vertical ($x–z$) plane with a slight undulation near the center of the array, as shown in Figure 6.3(b). The sensor depths vary from 200–210 m. The two acoustic sources are located at 500 m range from the center of the array, and are separated by 90° with respect the array; the source depth is 200 m (Figure 6.3(c) and (d)). The source positions are assumed to be known in x and y to within ±10 m, representing the approximate accuracy that could be obtained using DGPS (differential global positioning system) locations for the consort ships. The uncertainty in source depth is 2 m, consistent with the experiment described in [vB+89]. The reference hydrophone (mounted on the tow ship) is located 100 m to the left of the array (Figure 6.3(c)). The ocean is 4000 m deep with a typical N.E. Pacific sound-speed profile, shown in Figure 6.4.

The measured (synthetic) data consist of relative travel times along direct, bottom reflected, and bottom–surface reflected paths. The data were computed using the ray-tracing algorithm outlined in Section 6.2.4, and subsequently adding random (Gaussian) errors. Several different levels of error (standard deviation) are

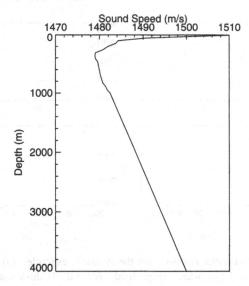

FIGURE 6.4. Ocean sound-speed profile for synthetic examples.

considered. In particular, the errors for the direct-path travel times are either 0.2, 0.5, or 1 ms. In each case, the errors for the bottom-reflected arrivals are twice as large as those on the direct arrivals, while the errors on the bottom-surface reflected arrivals are three times as large. The use of larger errors for reflected arrivals is designed to represent two factors that affect actual data measurements and inversion:

(i) the resolution in picking arrival times is generally lower for reflected paths due to the loss of high-frequency energy; and

(ii) modeling errors are larger for reflected arrivals due to the fact that the sea floor and surface are not completely flat and smooth.

In addition, the prior estimates for the source locations used in the inversion included Gaussian errors with standard deviations of 10 m for the x- and y-coordinates and 2 m for the z-coordinate. Finally, a sound-speed bias of standard deviation 2 m/s was included in the sound-speed profile used in the inversion.

Figures 6.5 and 6.6 illustrate the convergence properties of the inversion algorithm for data with an uncertainty of 0.2 ms (direct arrivals). Figure 6.5 shows the convergence in terms of the rms data misfit X, the rms change in sensor positions between iterations Δ, the rms misfit to the prior parameter estimates $\hat{X}$, and the array roughness R. The starting model (iteration 0) consists of a straight array (see Figure 6.6(a)) with the source positions corresponding to the prior estimates. Figure 6.5(a) shows that the data misfit X decreases by more than three orders of magnitude over the first two iterations, then remains constant at the desired value of $X = 1$ for iterations 3 and 4 while the rms change decreases below the threshold of $\Delta = 0.1$ m for convergence (Figure 6.5(b)). The rms prior misfit increases over the first two iterations from $\hat{X} = 0$ at the starting model to a value of 0.85

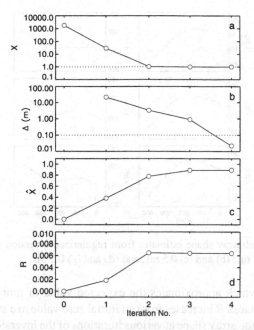

FIGURE 6.5. Convergence properties of the regularized inversion algorithm: (a) rms data misfit X; (b) rms sensor-position change between iterations Δ; (c) rms prior misfit $\hat{X}$; and (d) array roughness R.

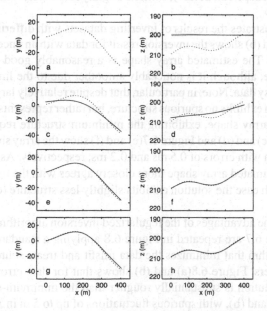

FIGURE 6.6. Towed-array shape at various iterations of the inversion: (a) and (b) show the starting model; (c) and (d) show iteration 1; (e) and (f) show iteration 2; and (g) and (h) show iteration 4 (final model). Dotted lines indicate the true array shape.

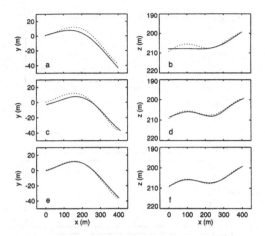

FIGURE 6.7. Towed-array shape estimates from regularized inversion for different error levels: (a) and (b) 1 ms; (b) and (c) 0.5 ms; and (d) and (e) 0.2 ms.

(Figure 6.5(c)), which approximates the expected value of unity. Similarly, the array-shape roughness R increases from an initial zero value to a stable final value. Figure 6.6 shows the array shape at various iterations of the inversion: it is apparent that the inversion algorithm introduces structure into the solution in a controlled manner. The final solution, shown in Figure 6.6(g) and (h), is an excellent estimate of the true array shape but, notably, exhibits slightly less overall curvature than the true shape.

Figure 6.7 illustrates the results of inverting datasets with differing error levels. Figure 6.7(a) and (b) shows the inversion result for data with an uncertainty of 1 ms (direct arrivals). The estimated array shape is a reasonably good approximation of the true shape, although it is noticeably smoother due to the limited resolving power of the noisy data. Note, in particular, that despite relatively large errors on the data, the solution exhibits no spurious structure, but rather represents a conservative estimate of the array shape, exhibiting the minimum structure required to fit the data. Figure 6.7(c) and (d) and Figure 6.7(e) and (f) show the array shapes estimated by inverting data with errors of 0.5 ms and 0.2 ms, respectively. As the data errors decrease, the estimated array shape more closely agrees with the true array shape. However, in each case the solution exhibits slightly less structure (curvature) than the true array.

To illustrate the advantages of the regularized inversion algorithm, the test cases shown in Figure 6.7 are repeated in Figure 6.8 applying a standard least-squares inversion algorithm that minimizes the data misfit and treats source positions as known parameters. Figure 6.8(a) and (b) shows that for data errors of 1 ms, the least-squares solution is substantially rougher than the minimum-structure result of Figure 6.7(a) and (b), with spurious fluctuations of up to 5 m in x and y, and up to 3 m in z. The fluctuations result from the tendency of a least-squares approach to over-fit the data, in effect fitting the noise as well as the data. In addition, the array-shape estimates in Figure 6.8 exhibit greater offsets from the true positions as a

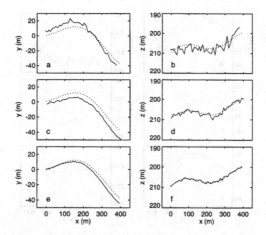

FIGURE 6.8. Towed-array shape estimates from least-squares inversion for different error levels: (a) and (b) 1 ms; (c) and (d) 0.5 ms; and (e) and (f) 0.2 ms.

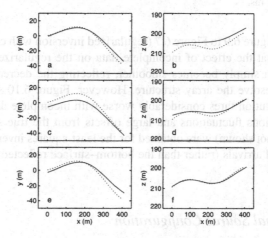

FIGURE 6.9. Towed-array shape estimates from regularized inversion using only direct and bottom-reflected paths for different error levels: (a) and (b) 1 ms; (c) and (d) 0.5 ms; and (e) and (f) 0.2 ms.

result of treating (erroneous) source positions as known parameters. Figure 6.8(c)–(f) show that the magnitude of the fluctuations decrease as the data errors decrease; however, the least-squares results remain significantly poorer than those of the regularized inversion.

The advantages of the regularized inversion are accentuated in cases where travel-time measurements are not available along all three ray paths. To illustrate this, Figures 6.9 and 6.10 show the regularized and least-squares results for the same data errors as in Figures 6.7 and 6.8; however, the inversion is applied to only the direct and bottom-reflected acoustic arrivals (i.e., the datasets do not include the bottom–surface reflected arrivals).

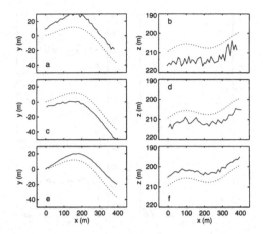

FIGURE 6.10. Towed-array shape estimates from least-squares inversion using only direct and bottom-reflected paths for different error levels: (a) and (b) 1 ms; (c) and (d) 0.5 ms; and (e) and (f) 0.2 ms.

Comparing Figure 6.9 to Figure 6.7 (regularized inversion with complete data), it is apparent that the effect of incomplete data on the regularized inversion is that the solution simply becomes smoother, reflecting the decreased ability of the dataset to resolve the array structure. However, Figure 6.10 shows that the least-squares solution fares considerably worse with incomplete data, exhibiting substantial spurious fluctuations and large offsets from the true solution. Even poorer results (not shown) were obtained for the least-squares inversion when the bottom-reflected arrivals (rather than the bottom–surface reflected arrivals) were omitted.

6.3.2 Optimal Source Configuration

In this subsection, the AEL error measure and optimization procedures developed in Section 6.2.3 are applied to investigate the effects of source geometry on towed-array shape estimation using the dual-shot method. The test case considered here is similar to that in Section 6.3.1, with a 400 m array towed at 200 m depth in 4000 m of water. However, since the optimization procedure is designed for a representative (expected) inverse problem, a straight array is assumed here. The data errors are 0.2 ms for the direct arrivals, 0.4 ms for bottom-reflected arrivals, and 0.6 ms for bottom–surface reflected arrivals. The source positions are assumed to be known to within 10 m in x and y and 2 m in z, and the potential sound-speed bias has a standard deviation of 2 m/s. A number of randomly generated trial inversions indicated that a trade-off parameter of $\mu_1 = 10$ is appropriate for this problem. The SSA algorithm described in Section 6.2.3 is applied to determine the source positions (relative to the array) that minimize the mean sensor-position error E given by (6.23).

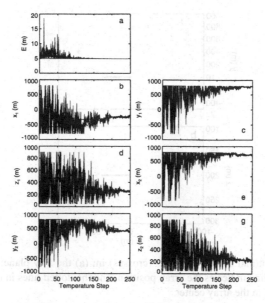

FIGURE 6.11. Convergence of the SSA optimization for source geometry: (a) shows the mean sensor-position error E; and (b)–(g) show the source coordinates $\{x_j, y_j, z_j, \ j = 1, 2\}$.

The convergence of the optimization procedure is illustrated in Figure 6.11, which shows the mean sensor-position error E and the source coordinates $\{x_j, y_j, z_j, \ j = 1, 2\}$ as a function of temperature step (all models in the simplex are shown). The search limits were set to be $(-800, 800)$ m for x and y, and $(100, 900)$ m for z. The annealing schedule for the optimizations was based on the requirement that ten model perturbations be accepted at each temperature step, with the temperature reduced by a factor of $\beta = 0.95$ between steps. Figure 6.11 shows that initially the source coordinates fluctuate over their entire allowed range, and sensor-position errors as large as $E = 19$ m are obtained. As the temperature is reduced, the error decreases steadily (although not monotonically), and the source coordinates gradually converge to fixed values. By temperature step 250, the error has essentially stopped decreasing, indicating that the various models in the simplex are simply fluctuating between good solutions. The optimization is then quenched to collapse the simplex to the single best model. The sensor-position error for the final solution is $E = 4.95$ m.

The optimal sensor configuration obtained by the SSA algorithm is shown in Figure 6.12. Figure 6.12(a) shows the source positions in the x–y plane; dotted lines indicating a 90° angle with respect to the array center are included as a reference. Note that, unlike the survey geometry suggested in [vB+89], the two sources are not equidistant from the array, and do not fall on the 90° lines. Figure 6.12(b) shows the source positions in the x–z plane. The sources are located at different depths, slightly deeper than the array. The optimal configuration in Figure 6.12 appears to be unique: repeating the SSA optimizations with different sequences, of random

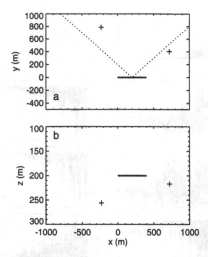

FIGURE 6.12. Optimal source positions (crosses) in: (a) the x–y plane; and (b) the x–z plane. The heavy line indicates the array position, and the dotted lines in (a) indicate a 90° angle with respect to the array center.

model perturbations produced configurations that were essentially identical to that shown, up to reflections about the x- and/or y-axes. However, Figure 6.11 indicates that in the course of the optimization, many different configurations were obtained that were *almost* as good as the optimal configuration. Therefore, it is interesting to compare the optimal configuration to the source geometry suggested in [vB⁺89] (sources at the same depth as array, equidistant from the array and separated by 90°). To this end, Figure 6.13 shows the mean sensor-position error E computed for the suggested configuration as a function of the range r from the sources to the array center. The error has a minimum of $E = 5.13$ m at a range $r = 339$ m. At shorter ranges, the error increases rapidly; at longer ranges, E increases gradually. The error $E = 4.95$ m obtained for the optimal configuration is included as a dotted line in Figure 6.13. It is apparent that the difference in the sensor-position error that results from using the optimal configuration or the suggested configuration

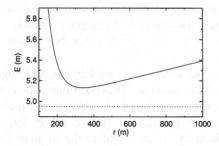

FIGURE 6.13. Mean sensor-position error E for the sources at the same depth as array, equidistant from the array and separated by 90° as a function of the range r from the sources to the array center. The dotted line indicates the error for the optimal configuration.

is small (for a good choice of r). In practice, the source geometry suggested in [vB+89] would seem to be an excellent choice.

6.4 Summary

This chapter considered the inverse problem of estimating the shape of a towed hydrophone array using the relative travel times of direct and reflected acoustic arrivals from sources deployed by a pair of consort ships (the dual-shot method). An algorithm was developed to invert dual-shot data for the most meaningful array-shape estimate. To date, this inversion has been solved as a least-squares problem (minimizing the squared data error), assuming straightline acoustic propagation and neglecting the inevitable errors in the source positions. The new approach is based on an iterated linearized inversion of the ray-tracing equations, which is solved using the method of regularization. The three-dimensional positions of both sources and sensors are treated as unknowns, subject to a priori information. For the sources, the prior information consists of position estimates and uncertainties. For the sensors, the prior information is that the array shape is expected to be smooth: this is applied by minimizing the three-dimensional curvature of the array to obtain a minimum-structure solution. The regularized inversion provides smooth solutions without the spurious fluctuations present in the least-squares solution. Fluctuations in the least-squares solution result from the tendency of a minimum-misfit approach to over-fit the data, in effect fitting the noise as well as the data. The regularized inversion avoids this by trading off data misfits with physical a priori information. In addition, treating source positions as (constrained) unknowns reduces offset errors in the solution, particularly in cases when not all acoustic arrivals can be used in the inversion.

Finally, the effect of the survey geometry was investigated by quantifying the sensor-position error using the a posterior uncertainty of the inversion. The optimal source configuration was determined by minimizing the sensor-position error with respect to the source positions using an efficient hybrid optimization algorithm. It was found that the standard source configuration typically employed for the dual-shot method, although nonoptimal, provides a good, practical approach to the acoustic survey.

Acknowledgment. This work was supported by a research contract with the Defence Research Establishment Atlantic, Dartmouth, NS, Canada.

6.5 References

[B+96]   G.H. Brooke, S.J. Kilistoff, and B.J. Sotirin. Array element localization algorithms for vertical line arrays, in J.S. Papadakis (ed.), *Proceedings of the Third European Conference on Underwater Acoustics*, pp. 537–542, Crete University Press, 1996.

[C⁺87]   S.C. Constable, R.L. Parker, and C.G. Constable. Occam's inversion: A practical algorithm for generating smooth models from electromagnetic sounding data, *Geophysics*, **52**:289–300, 1987.

[CD82]   K.C. Creager and L.M. Dorman. Location of instruments on the seafloor by joint adjustment of instrument and ship positions, *J. Geophys. Res.*, **87**:8379–8388, 1982.

[D⁺98a] S.E. Dosso, M.R. Fallat, B.J. Sotirin, and J.L. Newton. Array element localization for horizontal arrays via Occam's inversion, *J. Acoust. Soc. Am.*, **104**:846–859, 1998.

[D⁺98b] S.E. Dosso, G.H. Brooke, S.J. Kilistoff, B.J. Sotirin, V.K. Mcdonald, M.R. Fallat and N.E. Collison. High-precision array element localization of vertical line arrays in the Arctic Ocean", *IEEE J. Ocean. Eng.*, **23**:365–379, 1998.

[DS99]   S.E. Dosso and B.J. Sotirin. Optimal array element localization, *J. Acoust. Soc. Am.*, **106**:3445–3459, 1999.

[FD99]   M.R. Fallat and S.E. Dosso, Geoacoustic inversion via local, global and hybrid algorithms, *J. Acoust. Soc. Am.*, **105**:3219–3230, 1999.

[HB98]   M. Hansruedi and D.E. Boerner. Optimized and robust experimental design: a non-linear application of EM sounding, *Geophys. J. Int.*, **132**:458–468, 1998.

[HS94]   M. Hardt and F. Scherbaum. Design of optimum networks for aftershock recording, *Geophys. J. Int.*, **117**:716–726, 1994.

[H⁺96]   W.S. Hodgkiss, D.E. Ensberg, J.J. Murray, G.L. D'Spain, N.O. Booth, and P.W. Schey. Direct measurement and matched-field inversion approaches to array shape estimation, *IEEE J. Ocean. Eng.*, **21**:393–401, 1996.

[LT84]   L.R. Lines and S. Treitel. Tutorial: A review of least-squares inversion and its application to geophysical inverse problems, *Geophys. Prosp.*, **32**:159–186, 1984.

[L⁺95]   P. Liu, S. Hartzell, and W. Stephenson. Non-linear multiparameter inversion using a hybrid global search algorithm: Applications in reflection seismology, **Geophys. J. Int.**, **122**:991–1000, 1995.

[Mil83] P.H. Milne. *Underwater Acoustic Positioning Systems*. Cambridge University Press, Cambridge, 1983

[Old83] D.W. Oldenburg. Funnel functions in linear and nonlinear appraisal, *J. Geophys. Res.*: **88**:7387–7398, 1983.

[OC97]   J.C. Osler and D.M.F. Chapman. Seismo-acoustic determination of the shear-wave speed of surficial clay and silt sediments on the Scotian shelf, *Canad. Acoust.*, **24**:11–22, 1997.

[P⁺92]   W.H. Press, S.A. Teukolsky, W.T. Vetterling, and B.P. Flannery. *Numerical Recipes in FORTRAN*. Cambridge University Press, Cambridge, 1992.

[S⁺90]   J.A. Scales, P. Docherty, and A. Gersztenkorn. Regularisation of nonlinear inverse problems: Imaging the near-surface weathering layer, *Inverse Problems*, **6**:115–131, 1990.

[SH90]   B.J. Sotirin and J.A. Hildebrand. Acoustic navigation of a large-aperture array, *J. Acoust. Soc. Am.*, **87**:154–167, 1990.

[Ste76]   B.D. Steinberg. *Principles of Aperture and Array System Design*. Wiley, New York, 1976.

[SH76]   H. Szu and R. Hartley. Fast simulated annealing, *Phys. Lett. A.*, **122**:157–162, 1987.

[T⁺76]   W.M. Telford, L.P. Geldart, R.E. Sherif, and D.A. Keys. *Applied Geophysics*. Cambridge University Press, Cambridge, 1976.

[vB⁺89]   E.C. van Ballegooijen, G.W.M. van Mierlo, C. van Schooneveld, P.P.M. van der Zalm, A.T. Parsons, and N.H. Field. Measurement of towed array position, shape and attitude, *IEEE J. Ocean. Eng.*, **14**:375–383, 1989.

[vS89]   C. van Schooneveld. Inverse problems: A tutorial survey. In Y.T. Chan (ed.), *Underwater Acoustic Data Processing*, pp. 393–411, Kluwer Academic, Amsterdam, 1989.

[VH98]   H.T. Vincent II and S.-L.J. Hu. Geodetic position estimation of underwater acoustic sensors. *J. Acoust. Soc. Am.*, **102**:3099, 1998.

7

Mode-Coupling Effects in Acoustic Thermometry of the Arctic Ocean

Alexander N. Gavrilov
Peter N. Mikhalevsky

ABSTRACT The effects of mode coupling on modal travel times and amplitudes are considered with respect to the robustness and accuracy of acoustic thermometry in a range-dependent ocean waveguide. An approximate analytic solution for the complex amplitudes of coupled modes in a slowly varying waveguide is used to analyze the character of mode coupling effects. Change in the source–receiver transfer function for individual modes, due to the mode coupling, is discussed. It is shown that the variations of the modal travel times measured by the modal phases are much less sensitive to the mode coupling than those measured by locating the modal arrival time in pulse-like signals. The numerical solution of the coupled-mode problem has been used to model broadband acoustic propagation at 20 Hz over 1200 km from a source northwest of Franz Josef Land to a receiver in the Lincoln Sea, crossing the Eurasian continental slope, the deep-water Arctic Basin, and the Canadian continental slope. Year-round acoustic measurements are currently being made on this path as part of the Arctic Climate Observations using Underwater Sound (ACOUS) experiment that started in October 1998. The results of modeling show that the mode coupling limits the accuracy of acoustic thermometry on this path. However this limit should not exceed 5 millidegree C for integrated temperature changes on this path if the acoustic travel time variations are measured using the modal phase. Variations of the modal amplitudes due to mode coupling are examined using the results of the Transarctic Acoustic Propagation (TAP) experiment in 1994.

7.1   Introduction

The idea of acoustic thermometry in the ocean [MF89] is based upon the assumption that the individual acoustic modes, modal groups, or rays of the acoustic signals are stable and propagate along specific and known paths from the source to the receiver. Therefore, their travel times may be used as a measure of the range-averaged sound speed and variation and hence the water temperature and its variation along those paths. Such an assumption is acceptable, when the acoustic modes do not exchange their energy when propagating; i.e., the sound propagation is so-called mode-adiabatic. Solving the inverse problem of acoustic tomography and thermometry is much easier, if the sound propagation is mode-adiabatic, which makes it possible to associate each particular mode with its corresponding propagation path [Sh89]. However, because of range-dependent bathymetry and sound speed changes, the real ocean acoustic waveguide does not always satisfy the conditions of mode-adiabatic propagation. When such conditions do not exist, the acoustic modes are coupled with each other, which leads to a mixing of the modal energy and hence a much more difficult inversion of the acoustic thermometry results.

In acoustic tomography of oceanographic features, such as fronts or eddies, over relatively short paths, where acoustic parameters of the ocean waveguide are well investigated, the mode coupling effects may serve as an indication of changes in the ocean environment. The mode coupling effects due to the Barents Sea Polar Front have been theoretically studied and numerically modeled by Guoliang Jin et al. [JLCM96] using Evans' stepwise approach to the solution of the mode coupling problem [Ev83]. The travel time of coupled modes across the front was found to be different from that predicted using the adiabatic approximation. In acoustic thermometry for transoceanic paths with the parameters of the waveguide incompletely determined, the mode coupling leads to a poorly predictable perturbation of the signal parameters, which makes it difficult to perform oceanographic interpretation of the acoustic results. Numerical modeling for the waveform of the signals propagated over 1 Mm from ice camp Turpan to ice camp Narwhal in the TAP experiment [MBGS95], [MGB99], has shown that the mode coupling greatly influences the modal amplitudes and travel times [PFSO96].

As a part of the ACOUS project, year-round acoustic observations of long-term changes in the water temperature and sea ice thickness in the Arctic Ocean were started in October 1998. An acoustic source is moored in Franz Victoria Strait (FVS) northwest of Franz Josef Land (FJL) in water depth of 440 m suspended from the bottom to a depth of 60 m below the ice. A bottom moored autonomous receive array is deployed near the continental slope in the Lincoln Sea in a water depth of 545 m. This array will be recovered in the Fall of 2000 and will be replaced by another array to continue the recordings through October 2002 which is the projected life of the source. This acoustic path crosses the Central Arctic Basin with a depth exceeding 3.5 km. Both the Eurasian and Canadian continental slopes in the regions of FJL and the Lincoln Sea respectively are rather steep (up to 5°). Such a variable bathymetry along the path FVS–Lincoln Sea, as well as the horizontal gradient in the sound speed over the Eurasian continental slope results in noticeable coupling of the acoustic modes for low-frequency signals propagating over the path. The effects of mode coupling over the continental slopes may distort the signal structure and affect the main acoustic characteristics—modal travel times, phases, and amplitudes, which reflect changes in the water temperature and the ice thickness [GM95]. Therefore, it seems necessary to analyze the mode coupling effects, with respect to the robustness and the accuracy of both acoustic thermometry and ice monitoring over range-dependent paths in the Arctic Ocean.

In the first section of this chapter, we consider in general the problem of ocean acoustic thermometry in the presence of mode coupling. The approximate analytic solution for the complex coupled-mode amplitudes is considered to analyze the character of spatial variations in the modal amplitudes and phases due to mode coupling. The transfer function of coupled modes in an irregular waveguide is compared with that predicted with the adiabatic approximation.

The results of numerical modeling for the particular path from FVS to the Lincoln Sea are presented in the second section. The bathymetry data, the climatic sound speed profiles, and the ice statistics data along the path were used in the numerical modeling of the propagation for the calculation of the spatial variations of the complex modal amplitudes. The modeling has been performed for a broadband signal with the central frequency of 20 Hz. The influence of mode coupling on the modal amplitudes, phases, and travel times for this path is examined. The response of the signal parameters at the receiving site to simulated warming in the Atlantic water layer in the deep-water section of the path is also considered with particular attention to the mode-coupling effects.

In the third section, some of the TAP experiment results obtained in 1994, from the 1 Mm path from ice camp Turpan north of Svalbard to ice camp Narwhal in the Lincoln Sea, are considered with respect to mode-coupling effects and compared with the results of numerical modeling.

7.2   The Response of Acoustic Signal Parameters to Mode Coupling

To analyze the effects of mode coupling, we will represent the complex modal amplitudes Pn by a product of the rapidly varying exponential term in the adiabatic approximation and complex coefficients U_n which vary slowly with range r, as a result of mode coupling:

$$P_n(r, f) = U_n(r, f) \exp\{i\varphi_n(r, f)\}, \tag{7.1}$$

where

$$\varphi_n = \int_0^r k_n(r', f)\, dr', \tag{7.2}$$

and k_n are the modal wavenumbers, the imaginary parts of which account for the modal attenuation resulting mainly from sound absorption in the bottom sediments and acoustic scattering from the boundaries, including the sea ice in the Arctic. As a function of frequency, the modal amplitude $P_n(r, f)$ will be hereinafter referred as a modal transfer function. The two-scale decomposition of P_n has been argued by McDaniel [McD82] for solving the system of differential equations for coupled-mode amplitudes, which can be transformed into the system of differential equations for the complex envelopes U_n. This yields obvious advantages for numerical computing. For our study, the form of the solution given by (7.1) is particularly attractive, because it allows us to consider separately the influence of mode coupling on the modal amplitudes, phases, and travel times relative to the adiabatic-mode prediction of those parameters.

The ACOUS transArctic path crosses the continental slopes close to a perpendicular direction, so the two-dimensional approximation of low-frequency acoustic propagation along this path is reasonable. We will use the two-dimensional formulation of the coupled-mode approach thoroughly discussed in [CML96]. Omitting the discussion, let us consider the final equations. The sound pressure at distance r from an omnidirectional acoustic source in a waveguide is presented by a linear combination of normal modes

$$P(r, z, f) = \sum_n r_0 r^{-1/2} P_n(r, f) Z_n(r, z, f), \tag{7.3}$$

where Z_n are the orthogonal eigenfunctions of the differential operator

$$\mathbf{L} \equiv \frac{\partial^2}{\partial z^2} - \frac{1}{\rho(r, z)} \frac{\partial \rho(r, z)}{\partial z} \frac{\partial}{\partial z} + \left[k(r, z)^2 - k_n(r, f)^2 \right], \tag{7.4}$$

$$k(r, z) = 2\pi / c(z, r),$$

obeying the local boundary conditions, and the modal wavenumbers kn correspond to the eigenvalues of operator (7.4). If the modal amplitudes P_n are represented by (7.1), then the complex envelopes U_n of the modal amplitudes satisfy the system of differential equations

$$\left(\frac{d^2}{dr^2} + 2ik_n + \frac{d}{dr} + i\frac{dk_n}{dr} \right) U_n = \sum_m \left(A_{mn} \frac{d}{dr} + B_{mn} \right) U_m, \tag{7.5a}$$

$$A_{mn} = -\gamma_{mn} \exp\{i(\varphi_m - \varphi_n)\}, \tag{7.5b}$$

$$B_{mn} = \left(-i\gamma_{mn}k_m - \mu_{mn} \right) \exp\left[i\left(\varphi_m - \varphi_n \right) \right]. \tag{7.5c}$$

The coupling coefficients γ_{mn} and μ_{mn} satisfy the local boundary conditions. If the sea bottom consists of J sediment layers overlaying a flat, rigid basement, then the boundary conditions can be written as

$$Z_n(0, r, f) = 0,$$

$$Z_n\left(h_j^-, r, f\right) = Z_n\left(h_j^+, r, f\right), \qquad j = 1, 2, \ldots, J,$$

$$\frac{1}{\rho(r, h_j^-)} \frac{\partial}{\partial z} Z_n\left(h_j^-, r, f\right) = \frac{1}{\rho(r, h_j^+)} \frac{\partial}{\partial z} Z_n\left(h_j^+, r, f\right), \tag{7.6}$$

$$\frac{\partial}{\partial z} Z_n(h, r, f) = 0,$$

where h is the total thickness of the water column and the sediment layers, and $z = h_j(r)$ are the upper interfaces of the sediment layers. In (7.6) we assume that the upper boundary of the waveguide is a pressure–release surface, because, at such low frequencies as 20 Hz, the influence of Arctic ice on the eigenfunctions Z_n and the real parts of the wavenumbers k_n is negligible. Using boundary conditions (7.6) and the orthogonality of the eigenfunctions, one can derive the following expressions for the coupling coefficients [Fa92]:

$$\gamma_{mn} = \int_0^h \frac{2}{\rho} Z_n \frac{\partial Z_m}{\partial r} \, dr - \int_0^h \frac{1}{\rho^2} \frac{\partial \rho}{\partial r} Z_n Z_m \, dr$$

$$+ \sum_{j=1}^{J} \left[\left(\frac{1}{\rho} Z_n \frac{dh_j}{dr} Z_m \right) \Big|_{z=h_j^-} - \left(\frac{1}{\rho} Z_n \frac{dh_j}{dr} Z_m \right) \Big|_{z=h_j^+} \right],$$

$$\mu_{mn} = \int_0^h \frac{1}{\rho} Z_n \frac{\partial^2 Z_m}{\partial r^2} dr - \int_0^h \frac{1}{\rho^2} \frac{\partial \rho}{\partial r} Z_n \frac{\partial Z_m}{\partial r} dr$$

$$+ \sum_{j=1}^{J} \left[\left(\frac{1}{\rho} Z_n \frac{dh_j}{dr} \frac{\partial Z_m}{\partial r} \right) \Big|_{z=h_j^-} - \left(\frac{1}{\rho} Z_n \frac{dh_j}{dr} \frac{\partial Z_m}{\partial r} \right) \Big|_{z=h_j^+} \right], \quad (7.7)$$

Obviously, if the waveguide parameters vary smoothly with range, the coupling coefficients (mn are small, and the second term in (7.5c) is negligible relative to the first one even at very low frequencies.

The system of equations (7.5a–c) is usually solved by means of numerical integration of differential equations. Here we will consider the approximate solution of this system in an analytic form. Assume that each of the modes in the sound field is a superposition of the direct and backscattered waves propagating in opposite directions

$$P_n(r) = P_n^+(r) + P_n^-(r), \quad (7.8)$$

and the complex envelopes of those waves at arbitrary range $r = r_1$ are $U_n^+(r_1)$ and $U_n^-(r_1)$. The modal amplitudes $P_n^\pm(r)$ obey the system of equations (7.1.33) given in [BG]. Using the method of successive approximations, one can derive the solution of that system in the second approximation, which has the following form:

$$P_n^\pm(r) = U_n^\pm(r) \exp \left\{ \pm i \int_{r_i}^{r} k_n(r) dr' \right\}, \quad (7.9)$$

where the complex envelopes of the modes propagating in the positive direction from the source are

$$U_n^+(r) = U_n^+(r_1) \left[\frac{k_n(r_1)}{k_n(r)} \right]^{1/2}$$

$$+ k_n^{-1/2}(r) \sum_{m \neq n} U_m^+(r_1) \int_{r_1}^{r} b_{nm}^+(r') \left[\frac{k_n(r) k_m(r_1)}{k_m(r)} \right]^{1/2} \exp[i \, \Delta\varphi_{mn}(r')] \, dr',$$

$$(7.10)$$

where $\Delta\varphi_{mn}(r) = \int_{r_1}^{r} (k_m - k_n) \, dr'$. If the density ρ does not vary with range, i.e., $\partial\rho/\partial r = 0$, then $b_{nm}^+ = -\gamma_{mn}/2$. Integrating by parts in (7.10) and accepting $k_n(r)/k_m(r) \approx 1$, we obtain

$$U_n^+(r) = U_n^+(r_1) \left[\frac{k_n(r_1)}{k_n(r)} \right]$$

$$+ \frac{i k_n^{-1/2}(r)}{2} \sum_{m \neq n} k_m^{1/2}(r_1) U_m^+(r_1) \Big\{ \eta_{mn}(r) \exp[i \, \Delta\varphi_{mn}(r)] \Big|_{r_1}^{r}$$

$$- \int \frac{d\eta}{dr} \exp[i \, \Delta\varphi_{mn}(r)] \, dr' \Big\}, \quad (7.11)$$

where $\eta_{mn} = \gamma_{mn}/(k_m - k_n)$ is a small parameter characterizing the horizontal scale of spatial variations in the waveguide relative to the mode-to-mode interference cycles. If k_n and η_{mn} are infinitely differentiable functions of range, a progressive repetition of integration by parts produces the following expansion in (7.11):

$$U_n^+(r, f) = U_n^+(r_1, f)\left[\frac{k_n(r_1)}{k_n(r)}\right]^{1/2}$$
$$-\frac{i}{2}\sum_{m\neq n}\left[\frac{k_m(r_1)}{k_n(r)}\right]^{1/2} U_m^+(r_1, f)\{\chi_{mn}(r_1) - \chi_{mn}(r)\exp[i\,\Delta\varphi_{mn}(r, f)]\},$$

(7.12)

where

$$\chi_{mn}(r) = \sum_{l\geq 0} i^l\mathbf{L}^l\eta_{mn}(r),$$ (7.13)

$$\mathbf{L} \equiv \frac{1}{(k_m - k_n)}\frac{d}{dr}, \quad \text{and} \quad \mathbf{L}^0 \equiv 1.$$

Obviously, the series $i^l\mathbf{L}^l\eta_{mn}$ $(l = 0, 1, 2, \ldots)$ in (7.13) is converging, if the spectrum of horizontal variations of the waveguide parameters has a finite bandwidth $\xi < |k_m - k_n|$..

When the waveguide parameters vary with range slowly such that

$$\left|\frac{\dot{k}_n}{k_n(k_m - k_n)}\right| \ll 1 \quad \text{and} \quad \left|\frac{\dot{\gamma}_{mn}}{\gamma_{mn}(k_m - k_n)}\right| \ll 1,$$

the second term in the brackets in (7.11) is negligibly small relative to the first one, and therefore the approximate solution for $U_n^+(r)$ can be written as

$$U_n^+(r, f) = U_n^+(r_1, f)\left[\frac{k_n(r_1)}{k_n(r)}\right]^{1/2}$$
$$-\frac{i}{2}\sum_{m\neq n}\left[\frac{k_m(r_1)}{k_n(r)}\right]^{1/2} U_m^+(r_1, f)\{\eta_{mn}(r_1) - \eta_{mn}(r)\exp[i\,\Delta\varphi_{mn}(r, f)]\}.$$

(7.14)

Here, in (7.14) we specify the dependence of the complex modal amplitudes P_n and the differential phase Δ_{mn} on frequency. The modal wavenumbers k_n and the parameter η_{mn} vary with frequency much slower than P_n and $\Delta\varphi_{mn}$.

An equation similar to (7.14) can be derived from Eqs. 13 in [DCM86] by extracting the rapidly varying exponential term and ignoring the terms $O(\eta_{mn}^2)$. Certainly, expressions (7.12) and (7.14) cannot be used for accurate calculations of the sound field, if the waveguide is substantially irregular in the range direction. However, these expressions clearly show the character of spatial variations of the modal amplitudes and phase, due to mode coupling, and allow us to estimate the magnitude of those variations.

Assume that the waveguide is range dependent only within the distance from r_1 to r_2, and consider the modal amplitudes at $r > r_2$. If the waveguide parameters are smooth functions of range, then $\eta_{mn}(r_1) = \eta_{mn}(r_2) = 0$, and the second term dominates in the expansion in (7.12). Let an acoustic source be placed in a regular section of the waveguide at $r < r_1$. The approximate analytic solution (7.12) clearly shows that the spatial variations of the amplitude and the phase of any particular mode behind the irregular section depend on the amplitude and the phase of other modes, especially those with the adjacent numbers, at the beginning of the irregular section. This means that change in the propagation conditions in the regular waveguide at $r < r_1$ will lead to transformation of the functions $U_n(r)$ over the irregular section. Horizontal displacement of the source or change in the sound speed in the regular section, for instance, will cause variations of the modal amplitudes and a nonlinear response of the modal phases at a receiver located in the irregular section or behind it. Note that in the case of adiabatic-mode propagation the modal amplitudes at the receiver would remain constant, and the modal phases would depend linearly on the distance to the moving source and almost linearly on the sound speed.

The exponential terms in (7.12) and (7.14) vary rapidly with frequency. As a result, the magnitude-frequency characteristic of $U_n(r, f)$ is nonuniform, and the phase-frequency characteristic is nonlinear and may have both the positive and negative first derivative. Thus the modal transfer function (7.1) in an irregular waveguide is distorted by mode coupling, which may cause variations in the group delay of a broadband signal. Moreover, if the waveguide is irregular or partially irregular in the horizontal direction, the transfer function of the entire waveguide is not equal to the product of the transfer functions of its separate sections. In that sense, an irregular acoustic waveguide is a nonlinear transmission channel.

Monitoring of temporal changes in the ocean environment by means of low-frequency acoustics implies the measurements of relative change $\Lambda_n(t, f)$ in the modal transfer function for the acoustic transmission channel, which can be written as

$$
\Lambda_n(t, f) \equiv \frac{P_n(t_0 + t, f)}{P_n(t_0, f)}
$$
$$
= \frac{U_n(t_0 + t, f)}{U_n(t_0, f)} \exp\left\{ i \int_0^R \left[k_n(t_0 + t, f) - k_n(t_0, f) \right] dr \right\}, \quad (7.15)
$$

where R is a source-to-receiver distance. When the acoustic propagation is mode-adiabatic, the coefficient before the integral in (7.15) is equal to 1, and hence the phase variations of $\Lambda_n(f, t)$ in time are linearly connected with temporal changes in the real parts of the modal wavenumbers k_n, and the level of $|\Lambda_n(f, t)|$ correlates linearly with variations of the modal attenuation coefficients. However, if the modes in the acoustic signal are coupled with each other, then the coefficients $U_n(t)/U_n(t_0)$ become time dependent, and the relation between the functions $\Lambda_n(f, t)$ and the modal wavenumbers $k_n(t)$ is more complicated. Obviously, it is hardly probable to determine completely all the parameters of the real acoustic waveguide along the transoceanic path and calculate exactly the complex envelopes $U_n(r, f)$. Therefore, in the inverse problem, when restoring the temporal variations in the sound speed and the modal attenuation at a long and range-dependent acoustic path by tracking changes in the modal transfer functions, the term $U_n(t)/U_n(t_0)$ should be regarded

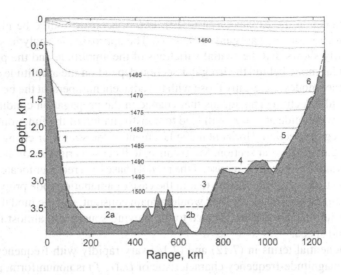

FIGURE 7.1. Sound-speed profile and bathymetry along the path FVS–Lincoln Sea: solid line—the ETOPO5 data (corrected for the Lincoln Sea); dashed line—a piecewise-linear approximation of bathymetry for acoustic modeling. Labeled regions: Eurasian continental slope (1); Nansen Basin (2a); Fram Basin (2b); region of Morris Jesup Plateau ((3) and (4)); and Canadian continental slope ((5) and (6)).

rather as random, multiplicative noise. Otherwise, it is necessary to measure the sound-speed distribution, the bathymetry, and the acoustic properties of the bottom along the path with a high accuracy.

7.3   Numerical Modeling for the Path FVS–Lincoln Sea

In this section we model the propagation of an acoustic thermometry signal on the path from FVS to the Lincoln Sea. We assume that an acoustic source is located at the edge of the Eurasian continental shelf in FVS northwest of FJL, and a receive array is deployed at the edge of the Canadian continental shelf in the Lincoln Sea. The acoustic thermometry signal is assumed to be broadband with a center frequency of 20 Hz.

The ETOPO5 database and the bathymetry charts of the Lincoln Sea were used to determine the bottom profile along the path FVS—Lincoln Sea. The ETOPO5 data may be insufficiently accurate for the region of the Lincoln Sea, because they show the Canadian continental slope to be much more gradual than that shown in the bathymetry charts commonly accepted [AOF71], [AOW83]. For numerical modeling, the bottom profile along the path, shown in Figure 7.1, was approximated by six linear sections.

The oceanographic data along the path were taken from the AARI Arctic climatology database [ACA96]. The sound speed was computed by the Chen–Millero–Li

TABLE 7.1. Geoacoustic parameters of the sediment layers used in the model of the bottom.

	Layer 1	Layer 2
Depth below the bottom surface, m	0–50	50–4500[a]
Compressional wave speed at the upper interface, m/s	1800	2000
Gradient of compressional wave speed, s^{-1}	2	1
Compressional wave attenuation, dB/m/kHz	0.04	0.2
Shear wave speed at the upper interface, m/s	150	300
Gradient of shear wave speed, s^{-1}	3	1
Shear wave attenuation, dB/m/kHz	0.15	0.2
Density, kg/m^3	2100	2300

[a] Below sea level.

formula [ML94]. A total of five different sound speed profiles were specified at seven locations along the path. The sound speed field in both the Nansen and Fram Basins was assumed to be horizontally regular along the path [see Figure 7.1]. The sound-speed profiles between the specified locations were calculated by linear interpolation to points along an equidistant grid with a 10 km horizontal step.

The acoustic model of the bottom consists of two layers of sediments overlaying the rigid basement. The depth of the basement interface was set at 4500 m below the sea level. The upper layer was assumed to have a constant thickness of 50 m along the entire path. The geoacoustic parameters of the sediment layers, given in Table 7.1, were also assumed to be invariant along the path in the model. These parameters are, on the one hand, in reasonable agreement with the geoacoustic model of the Lincoln Sea [Ge90], and, on the other hand, do not contradict the geoacoustic data available for FVS. Clearly, such a simple and invariant model cannot suit the real geoacoustic characteristics over the whole path. However, bottom interaction in the deep-water section of the path influences only the high-order modes that attenuate rapidly over the continental slope and play a negligible role in the received signal.

For the modeling, the number of normal modes was limited to 20, which is less than the number of modes propagating in the deep-water section, but twice the number of modes ducted within the sea surface and the bottom in the shallow-water sections. This allowed for coupling from the high-order modes to those of lower order over the continental slopes.

The sound scattering from the Arctic ice pack was included in the acoustic propagation model only for the calculation of acoustic loss at the upper boundary of the waveguide. We did not compute the scattered sound field. According to mapping of the ice draft and roughness in the Arctic Ocean [BM92], the statistics parameters in the model of the ice cover were chosen to be range dependent along the path. The acoustic reflection coefficient for the rough ice cover was calculated with the use of the Arctic ice scattering model recently developed by Kudryashov [Ku96], [GK94]. This full elastic model uses separate statistics for level ice and ice ridges. Such two-scale ice statistics have been provided for the central part of the Arctic Basin by processing the data of ice draft measurements from submarines [BAZ94]. To define the two-scale statistics parameters for the whole path, we assumed that considerable increase in the mean ice draft and its standard deviation north of Greenland and the Canadian Archipelago was the result of both the higher concentration of ridged ice and the greater thickness of ice in general. The ice statistics parameters used in the model are given in Table 7.2. The acoustic characteristics of sea ice were taken as follows: compressional wave speed, 3500 m/s; shear wave speed, 1800 m/s; compressional wave attenuation, 0.15 dB/m/kHz; and shear wave attenuation, 1.0 dB/m/kHz; and density, 910 kg/m^3.

The computer code for numerical modeling of the signal propagation at the path FVS–Lincoln Sea starts with the evaluation of the modal eigenfunctions $Z_n(z)$ and the complex wavenumbers k_n at the points spaced at 10 km intervals along the path. The modal eigenfunctions and the wavenumber were calculated using the computer code KRAKEN-C [Po91] with the angular dependence of the ice reflection coefficient in the code's input, computed via Kudryashov's algorithm. The derivatives of the eigenfunctions and the wavenumbers with respect to range were computed at those points by finite-difference methods. To calculate the coupling coefficients, the integrals in (7.7) were evaluated by regular methods of numerical integration.

To evaluate the system of differential equations (7.5a–c), we used a fourth-order Runge–Kutta method. The range step size of the numerical integration was chosen to provide a nondiverging solution within each section of the path with the minimum computation time. The step size of 150 m provided a relative accuracy of the solution not worse than 10^{-4}. The range dependence of the wavenumbers k_n and the coupling coefficients γ_{mn} was approximated by stepwise-linear interpolation

TABLE 7.2. Ice statistics parameters along the path FVS–Lincoln Sea used in the acoustic propagation model.

	Sections 1–3	Section 4	Sections 5–6
Range, km	0–790	790–1020	1020–1250
Level ice:			
Proportion	0.5	0.3	0.2
Mean draft, m	3.0	3.5	4
Standard deviation, m	1.0	1.2	1.5
Correlation length	1000[a]	1000	1000
Ridged ice:			
Proportion	0.5	0.7	0.8
Mean draft, m	5.0	5.8	6.5
Standard deviation, m	3.0	3.5	4.2
Correlation length, m	40	35	35
Upper-to-lower roughness STD ratio	1/4	1/4	1/4
Upper-to-lower roughness correlation	0.7	0.7	0.7

[a] The correlation length of level ice roughness varies in a wide range from hundreds of meters to several kilometers.

of the values computed on the 10-km grid. The coupling coefficients μ_{mn} were found to be small and, therefore, ignored in calculations.

To define the initial condition $U_n(0)$ and $U_n'(0)$ at the source, a short, range-independent section can be formally introduced at the beginning of the path. Then, if neglecting backscattering, the initial conditions for the first irregular section are as follows:

$$
U_n(0) = \frac{1}{\rho(0, z_0)} \left[\frac{2\pi}{k_n(0)} \right]^{1/2} \exp(i\pi/4) Z_n(0, z_0),
$$

$$
\frac{d}{dr} U_n(0) = 0,
$$

(7.16)

where z_0 is the depth of the source. For numerical calculations, the source depth was assumed to be 100 m.

As a result of the complete algorithm, the complex envelopes $U_n(r)$ and the modal amplitudes $P_n(r)$ were obtained as functions of range. To calculate the modal transfer function for a broadband signal, the full procedure described above was performed at frequencies from 19 Hz up to 21 Hz with a step of 0.1 Hz. Such a step size was found to be small enough for the coupling coefficients to vary gradually with the frequency increment. The transmitted signal was modeled by a Gaussian pulse modulating the carrier of 20 Hz. The pulse width was assumed to be 1 s, which corresponded to the spectrum bandwidth of ±1 Hz.

The spatial variations of the modal amplitudes and phases, due to mode coupling at the first 100 km section of the path, are shown in Figures 7.2(a) and 7.2(b) respectively. These variations decrease rapidly with the increase of sea depth, because, in deeper water, the lower-order modes are trapped in the upward refracting water channel, and therefore do not interact with the bottom. Small, long-period fluctuations in the complex envelopes of the lower-order modes (especially modes 1 and 2) over the deep-water part of the section are a result of a gradual increase in the thickness of the Arctic near-surface acoustic duct over the Eurasian continental slope. Such small fluctuations cause insignificant divergence of the modal energies from their adiabatic values. Thus the contribution of the variations in the sound speed along the propagation path to the mode coupling is negligible relative to that of the range-dependent bathymetry.

Figure 7.3 shows the spatial variations of the modal complex envelopes (magnitude (a) and phase (b)) along the last 100 km section of the path FVS–Lincoln Sea. Here, the sound speed is less variable in the horizontal direction, and hence the amplitudes of the low-order modes are almost uniform in the deep-water region. The coupled-mode variations become considerable over the last, most abrupt 30 km part of the continental slope. Because of upward refraction in the Arctic channel, a normal mode of lower order propagates closer to the sea surface than that of a higher order. Therefore, the influence of bottom interaction and mode coupling on different modes becomes noticeable at different sea depths, which is clearly seen in Figure 7.3.

The magnitude-frequency and phase-frequency characteristics of the complex modal amplitude $U_n(f, R)$ at the receiving site are shown in Figures 7.4(a) and

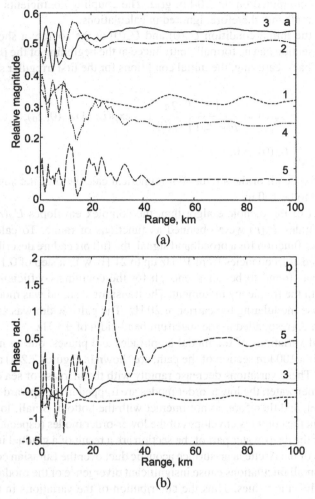

(a)

(b)

FIGURE 7.2. Spatial variation of (a) the magnitude and (b) the phase of the complex envelopes $U_n(r)$ at the first 100 km section of the path FVS–Lincoln Sea. Curves are labeled with the mode numbers.

7.4(b), respectively. The modal amplitudes vary considerably in a relatively narrow frequency band of 2 Hz. The perturbation of the modal phases due to mode coupling is nonlinear in frequency. Both the magnitude and the phase of mode 1 fluctuate rapidly with frequency, while the other modes vary gradually. This is so, because the difference in the wavenumber derivative with respect to the frequency for mode 1 and mode 2 is considerably greater than that for the other modes with the adjacent numbers. This also follows from the approximate solution (7.12). Since the coupling coefficients and the parameter η_{mn} decrease rapidly with the increase of the difference $(m - n)$, the terms corresponding to the adjacent modes dominate the other terms in the sum in (7.12). Assuming that the bandwidth of the signal

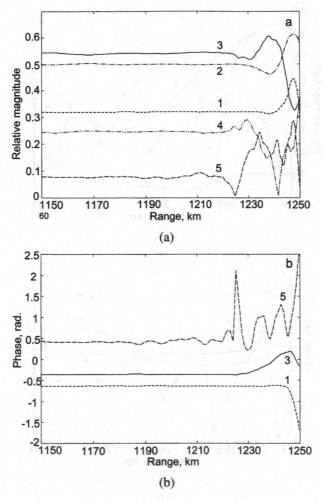

(a)

(b)

FIGURE 7.3. Same as in Figure 7.2, but for the last 100 km section of the path.

spectrum is relatively small, and taking into account that the mode coupling is significant only at the beginning and end of the path, (7.12) can be written in the following approximate form:

$$U_n(R, \omega) \approx U_n(0, \omega) - \frac{i}{2} \sum_{m=n-1, n+1} U_m(0, \omega_0)$$

$$\times \left(\chi_{mn}(0, \omega_0) - \chi_{mn}(R, \omega_0) \exp \left\{ i R \left[(\bar{k}_m - \bar{k}_n) + \Delta\omega (\bar{k}'_m - \bar{k}'_n) \right] \right\} \right),$$

$$(7.17)$$

where

$$\bar{k}_n = \frac{1}{R} \int_0^R k_n(r, \omega_0) \, dr, \qquad \bar{k}'_n = \frac{1}{R} \int_0^R \frac{\partial}{\partial \omega} k_n(r, \omega) \Big|_{\omega=\omega_0} dr, \qquad \Delta\omega = \omega - \omega_0,$$

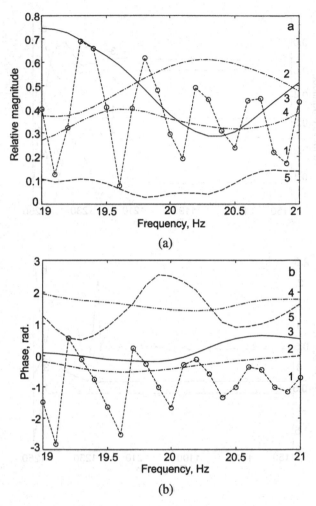

FIGURE 7.4. (a) Magnitude-frequency and (b) phase-frequency characteristics of the modal amplitudes $U_n(f)$ at the receiving site of the path FVS–Lincoln Sea.

and $\omega_0 = 2\pi f_0$ is the center frequency. Expression (7.17) clearly shows that the transfer function of mode 1 should be modulated with the approximate period

$$F_M \approx \left|1/r\left(\bar{k}_2' - \bar{k}_1'\right)\right| \approx \left|1/(t_2 - t_1)\right|,$$

where t_1 and t_2 are the travel times of modes 1 and 2, respectively. The difference in the travel time of modes 1 and 2 at the path FVS–Lincoln Sea is about 7.5 s, while that for the other pairs of modes is less than 1 s (see Figure 7.5). The influence of mode 1 on mode 2 is much weaker, because it has a much smaller amplitude.

Modulation in the transfer function leads to splitting of the modal pulse.

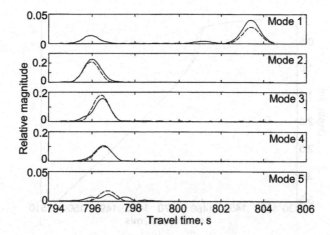

FIGURE 7.5. The arrival pattern of modes 1–5 in a 1 s pulse signal propagated over the path FVS–Lincoln Sea (solid line, coupled-mode model; dashed line, adiabatic-mode model).

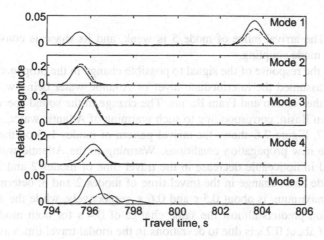

FIGURE 7.6. Same as in Figure 7.5, but for the case of hypothetical warming by 0.2° C in the Atlantic water layer over the path's section in the Nansen and Fram Basins (section 2 in Figure 7.1).

Additional peaks in the modal arrival pattern (envelope of the signal amplitude as a function of time) appear with the time delay $1/F_M$ relative to the main modal arrival. Such additional pulses are generated by intermodal energy transport due to mode coupling over the continental slope. Figure 7.5 shows the arrival pattern of modes 1–5 on the receive array (solid lines). Here, one can see two prominent arrivals of mode 1: the later and higher arrival peak coincides with that predicted by adiabatic approximation (dashed line), while the first peak arrives with mode 2. Modes 2, 3, and 4 have single arrival peaks that almost coincide with the adiabatic

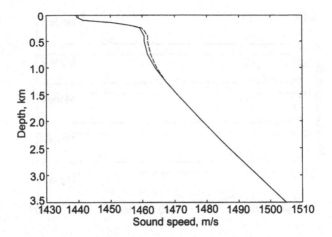

FIGURE 7.7. The sound-speed profile in the Nansen Basin used for acoustic modeling (solid line), and the same profile modified for assumed warming by 0.2° C in the intermediate layer of Atlantic water (dashed line).

prediction. The arrival pulse of mode 5 is weak, and its shape is considerably distorted by mode coupling.

To model the response of the signal to possible change in the propagation conditions, we assumed the intermediate layer of Atlantic water being warmer by 0.2° C over the Nansen and Fram Basins. The change in the sound speed profile in the Nansen Basin, corresponding to such warming of Atlantic water, is shown in Figure 7.7. Figure 7.6 shows the arrival pattern of modes 1–5 on the receive array for the new propagation conditions. Warming in the Atlantic water layer is responded in noticeable decrease in the travel time of modes 2 and 3. In the coupled-mode case, change in the travel time of modes 2 and 3, determined by the peak's maximum, is about 0.5 s and 0.6 s, respectively, while the adiabatic approximation predicts almost the same change of 0.4 s for both modes. This difference of about 0.2 s is due to deviations in the modal travel times as a result of mode coupling. Note that the amplitude of the coupled modes also fluctuates, which is clearly seen when comparing Figure 7.5 and 7.6. The amplitude variation of modes 2 and 3 is about 30%, i.e., 2 dB. The pulse shape of the higher-order modes (mode 5 and higher) may be subjected to substantial distortion resulting from mode coupling.

The origin of such fluctuations becomes clearer after comparing the frequency characteristics of $U_n(f, R)$ for the two modeled propagation conditions. Figure 7.8 shows the magnitude and phase characteristics of $U_n(f, R)$, as those in Figure 7.4, but for the modified propagation conditions. Change in the sound speed in the regular, deep-water section of the path has led to transformation of the functions $U_n(f, R)$, which has resulted in additional variations in the modal travel times and amplitudes.

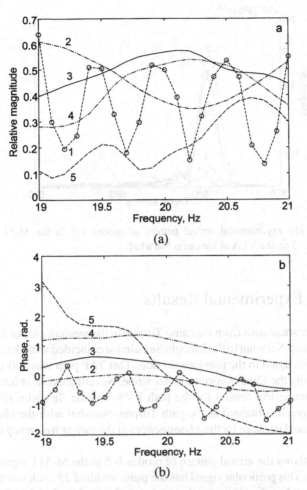

(a)

(b)

FIGURE 7.8. Same as in Figure 7.4, but for the warmed Atlantic water layer in the Nansen and Fram Basins.

Note that the phase variations of modes 2 and 3 due to mode coupling do not exceed 1 radian, which corresponds to about 8 ms in terms of the travel time. On the other hand, a linear phase variation of 1 radian within the 2 Hz frequency band of the propagated pulse leads to a shift in the location of the modal pulse maximum of 80 ms. Clearly, changing phase slopes within the frequency band due to mode coupling will lead to pulse shape distortion and temporal dispersion of the pulse, making travel time estimates based upon detecting the pulse peak subject to potentially large errors. Regarding acoustic thermometry, this means that the travel time measurements, by tracking the modal phase directly, are less sensitive to mode coupling than the measurements of the modal arrival time by detecting the peak of the pulse in time.

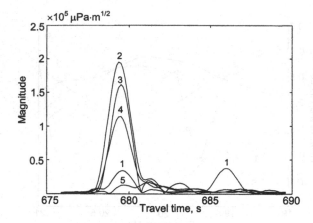

FIGURE 7.9. The experimental arrival pattern of modes 1-5 in the M-511 signal of transmission No.5 on the VLA at ice camp Narwhal.

7.4   The Experimental Results

The TAP experiment path from ice camp Turpan in the western part of the Nansen Basin to ice camp Narwhal in the Lincoln Sea almost coincided with the path FVS–Lincoln Sea discussed in the previous section. That TAP path was 300 km shorter and crossed only the Canadian continental slope. Nevertheless, it is interesting to compare the numerical results for the path FVS–Lincoln Sea with some of the experimental results obtained at the path Turpan–Narwhal with the phase-coded signals modulated in phase by the M-sequences at the carrier frequency of 19.6 Hz [GA96].

Figure 7.9 shows the arrival pattern of modes 1–5 in the M-511 signal of transmission No. 5. This particular signal has the pulse width of 25 cycles of the carrier, i.e., approximately 1.3 s. The experimental waveform of mode 1 has a secondary pulse arrived with the higher-order modes, as well as that numerically simulated for the path FVS–Lincoln Sea.

The distributions of the modal amplitudes in four different transmissions (Nos. 5, 13, 19, and 31) separated in time with intervals of 20, 39, and 25 hr, are shown in Figure 7.10. The modal distributions exhibit considerable fluctuations in the amplitudes of mode 3 and higher. Such fluctuations cannot be explained by adiabatic mode theory. Indeed, the environmental conditions at Turpan and Narwhal did not change significantly during the experiment. Moreover, Narwhal drifted very slowly—3 km in the first 4 days of the experiment, and hence the bathymetry and the acoustic properties of the bottom in the shallow-water part of the path could not change considerably. The main change in the propagation conditions was related to variations of the source-to-receiver distance due to the relatively fast drift of Turpan. Thus, the modal amplitudes would have been stable in the experiment, if the acoustic propagation from Turpan to Narwhal was mode-adiabatic.

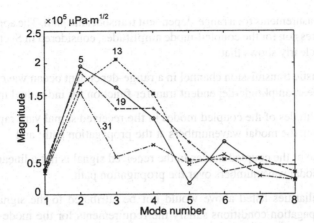

FIGURE 7.10. Magnitude of the arrival peaks of modes 1–8 filtered on the VLA at Narwhal in transmissions Nos. 5, 13, 19, and 31.

Different causes of these mode fluctuations in the TAP signals were investigated, including possible "spurious mode coupling" due to imperfect filtering of the modes on the vertical array at Narwhal. These fluctuations may also be the result of experimental errors and environmental uncertainty, such as ambient noise, uncompensated tilt of the array, insufficient number of receivers in the array, or imperfect determination of the modal eigenfunctions. However, when processing the TAP signals, these errors were eliminated or considerably diminished [GA+96]. The sound speed profile was measured at Narwhal several times during the experiment, and did not exhibit significant variations in the environmental conditions. Knowing the real sound-speed profile and the geoacoustic parameters of the Lincoln Sea provided good data for accurate modeling of the modal eigenfunctions. The array tilt in transmissions Nos. 5, 13, and 19 was known with a sufficient accuracy by the array element localization system. In transmission No. 31, the array tilt in the source-to-receiver direction was estimated by the vertical phase profile of mode 1 separated in time from the other mode after correlation with the signal replica. As a result, the maximum relative error in the estimates of the modal amplitudes is found not to exceed a few percent. This means that such "spurious" mode-coupling effects could not produce the observed modal fluctuations shown in Figure 7.10. Also, the appearance of the secondary peak in the arrival pattern of mode 1 cannot be due to imperfect mode filtering. Thus, the main cause of modal fluctuations in the TAP signals was the mode coupling over the Canadian continental slope, as a result of the variations in the source-to-receiver distance.

7.5   Conclusions

In this chapter, the influence of mode coupling on the modal amplitudes, phases, and travel times is considered with respect to the robustness of acoustic thermometry measurements for a range-dependent transoceanic path. The approximate analytic expression for the coupled-mode amplitudes, considered in Section 7.1 of the chapter, clearly shows that:

(1) the acoustic transmission channel in a range-dependent ocean waveguide has a nonlinear, amplitude-dependent transfer function for individual modes;

(2) the amplitudes of the coupled modes in the received signal vary rapidly with changes in the modal wavenumbers at the propagation path; and

(3) the phase of the coupled modes in the received signal is not a linear function of the modal wavenumbers over the propagation path.

All the peculiarities listed above would not be attributed to the signals, if the acoustic propagation conditions satisfy the requirements for the mode-adiabatic approximation.

A numerical algorithm and a computer code for the calculation of the modal transfer function and the arrival pattern in a range-dependent ocean waveguide have been developed. The propagation of a broadband signal with a central frequency of 20 Hz and a bandwidth of 2 Hz at the transArctic path FVS–Lincoln Sea has been modeled. The results of modeling for the coupled-mode propagation are compared with those predicted from the mode-adiabatic approximation. The response of the modal transfer functions and the modal pulse shape and arrival time to hypothetical warming of the Atlantic water layer by $0.2°$ C over the path's sections in the Nansen and Fram Basins is analyzed. It is found that the mode-coupling results in considerable fluctuations of the modal amplitudes, and leads to additional deviation in the modal travel times which does not correlate linearly with the change in the sound speed. The relative variations in the amplitudes of modes 2 and 3, the least sensitive to the mode-coupling effects, reach up to 30%. This may render it more difficult to monitor changes in the mean Arctic ice thickness over the range-dependent path by way of measuring long-term variations in the modal propagation loss. The deviation of the modal phases and travel times due to mode coupling will limit the resolution of acoustic thermometry at the path FVS–Lincoln Sea. However, the acoustic travel time measurements by the modal phase are much less sensitive to mode-coupling effects than those by the modal arrival time detected by locating the peak of the modal pulse. Thus, using the modal phase, as an acoustic travel time measure, is more robust for acoustic thermometry in the ocean, when the mode-adiabatic approximation is used for solving the inverse problem, while the actual acoustic propagation conditions are not mode-adiabatic. The coupled-mode limitation for the travel time measurements by the modal phase is less than 10 ms for the path FVS–Lincoln Sea, i.e., only 5 millidegree in terms of path-averaged temperature (see [PFSO96]).

Acknowledgments. We are thankful to Dr. Mikhail Andreyev the of General Physics Institute of the Russian Academy of Sciences, who processed most of the signals received at Narwhal during the TAP experiment. We also thank Dr.

Vladimir Kudryashov of the N.N. Andreyev Acoustics Institute for assistance in programming the algorithm for numerical calculation of the acoustic ice scattering. This work was carried out under the sponsorship of the Office of Naval Research and the Russian Ministry of Science and Technology.

7.6   References

[ACA96]   *Arctic Climatology Atlas.* on CD-ROM: AARI, St. Petersburg, http://ns.noaa.gov/atlas/, 1996.

[AOF71]   *Arctic Ocean Floor* (chart). (C. Bruce and T. Marie (ed.)), National Geographic Society, New York, 1971.

[AOW83]   *Arctic Ocean of World Ocean Atlas*, Vol. III (S.G. Gorshkov, (ed.)), Pergamon, New York, 1983.

[BAZ94]   S.V. Brestkin, E.O. Akseonov, and V.F. Zakharov. Statistical characteristics of the bottom profile of the Arctic ice cover. Tech. Rep. 94/3, AcoustInform, Moscow, Russia, May 1994.

[BG]   L.M. Brekhovskikh and O.A. Godin. *Acoustics of Layered Media, 2: Point Sources and Bounded Beams.* Springer Series on Wave Phenomena, Vol. 10. Springer-Verlag, Berlin, 1992.

[BM92]   R.H. Bourke and A.S. McLaren. Contour mapping of Arctic Basin ice draft and roughness parameters. *J. Geophys. Res.*, **97**(C11): 17715–17728, 1992.

[CML96]   C-S. Chiu, J.H. Miller, and J.F. Lynch. Forward coupled-mode propagation modeling for coastal acoustic tomography. *J. Acoust. Soc. Am.*, **99**:793–802, 1996.

[DCM86]   Y. Desaubies, C.-S. Chiu, and J.H. Miller. Acoustic mode propagation in a range-dependent ocean. *J. Acoust. Soc. Am.*, **80**:1148–1160, 1986.

[Ev83]   R.B. Evans. A coupled mode solution for acoustic propagation in a waveguide with stepwise depth variations of a penetrable bottom. *J. Acoust. Soc. Am.*, **74**:188–195, 1983.

[Fa92]   J.A. Fawcett. A derivation of the differential equations of coupled-mode propagation. *J. Acoust. Soc. Am.*, **92**:290–295, 1992.

[GA96]   A.N. Gavrilov and M.Yu. Andreyev. Analysis of the results of the Transarctic Acoustic Propagation experiment. Tech. Rep. No. 2/96, Marine Science Int. Corp., North Falmouth, MA, 1996.

[GA+96]   A.N. Gavrilov and M.Yu. Andreyev. New results of the TAP experiment data analysis. In Tech. Rep. "Large scale variations in the Arctic Ocean and the acoustic monitoring scheme optimal for observations of climatic changes," A.N. Gavrilov (ed.). Tech. Rep. No. 5/96, Marine Science Int. Corp., North Falmouth, MA, pp. 24–32, 1996.

[Ge90]   W.H. Geddes. Geoacoustic model of the Lincoln Sea. GGAI Tech. Rep. 3-90, March 1990.

[GK94]   A.N. Gavrilov and V.N. Kudryashov. Numerical modeling of low-frequency sound propagation in a horizontally stratified Arctic waveguide. Tech. Rep. 94/1, AcoustInform, Moscow, Russia, January 1994.

[GM95]   A.N. Gavrilov and P.N. Mikhalevsky. Modeling an acoustic response to long-term variations of water and ice characteristics in the Arctic Ocean, in *Proceedings of Oceans '95 IEEE Conf.*, San Diego, CA, Vol. 1, pp. 247–253, June 1995.

[JLCM96] G. Jin, J.F. Lynch, C.-S. Chiu, and J.H. Miller. A theoretical and simulation study of acoustic normal mode coupling effects due to the Barents Sea Polar Front, with application to acoustic tomography and match field processing. *J. Acoust. Soc. Am.*, **100**:193–206, 1996.

[Ku96]   V.M. Kudryashov. Calculation of the acoustic field in an Arctic waveguide. *Phys. Acoust.* **42**:386–389, 1996.

[MBGS95] P.N. Mikhalevsky, A.B. Baggeroer, A.N. Gavrilov, and M.M. Slavinsky. Experiment tests use of acoustics to monitor temperature and ice in Arctic Ocean. *EOS*, **6**(27):265–269, 1995.

[McD82]   S.T. McDaniel, Mode coupling due to interaction with the seabed. *J. Acoust. Soc. Am.*, **72**:916–923, 1982.

[MGB99]   P.N. Mikhalevsky, A.N. Gavrilov, and A.B. Baggeroer. The transarctic acoustic propagation experiment and climate monitoring in the Arctic. *IEEE J. Ocean. Eng.*, **24**(2):183–201, 1999.

[ML94]   F.J. Millero and X. Li. Comments on "On equations for the speed of sound in sea water." *J. Acoust. Soc. Am.*, **95**:2757–2759, 1994.

[MF89]   W. Munk and A.M.G. Forbes. Global ocean warming: An acoustic measure. *J. Phys. Oceanogr.*, **19**:1765–1778, 1989.

[PFSO96] R. Pawlowicz, D. Farmer, B. Sotirin, and S. Ozard. Shallow-water receptions from the transarctic acoustic propagation experiment. *J. Acoust. Soc. Am.*, **100**:1482–1492, 1996.

[Po91]   M.B. Porter. The KRAKEN normal mode program. Tech. Rep., SACLANT Undersea Research Center Mem. (SM-245)/ NRL Mem. Rep. 6920, 1991.

[Sh89]   E. C. Shang. Ocean acoustic tomography based on adiabatic mode theory. *J. Acoust. Soc. Am.*, **85**:1531–1537, 1989.

8

On the Characterization of Objects in Shallow Water Using Rigorous Inversion Methods

Bernard Duchêne
Marc Lambert
Dominique Lesselier

ABSTRACT We are concerned herein with inverse obstacle scattering problems in underwater acoustics, where the goal is to characterize an unknown object from measurements of the pressure field which results from its interaction with a known probing (incident) wave. Two configurations are considered, i.e., an impenetrable, sound-soft or sound-hard object immersed in a shallow-water open waveguide, the source and the receivers also being located in it, and a penetrable object embedded in a semi-infinite sediment, illuminated and observed from a semi-infinite water column. The inverse problem consists in retrieving the contour of the impenetrable object or a contrast function representative of the constitutive physical parameters of the penetrable one. This is done by means of deterministic nonlinearized iterative solution methods, one devoted to each configuration, i.e., the distributed source method and the binary modified gradient method. Both of them attempt to build up a solution by minimizing, in an appropriate L_2 setting, a two-term cost functional which expresses the discrepancies between the fields computed by means of the retrieved solution and the data, the latter being either the field measured on the receivers or the known incident field on the boundary of the object (impenetrable case) or inside it (penetrable case). In both configurations the well-known ill-posedness of the inverse scattering problem is enhanced either by range filtering or by the limited aspect of the data, a strong regularization being then needed. This is done by introducing, in the inversion algorithms, some a priori information on the object to be retrieved, which consists in the smoothness of its contour or in its homogeneity.

8.1 Introduction

The Electromagnetic Research Department (DRE/CNRS-Supélec) has been involved for a long time in inverse problems both in electromagnetics and in acoustics. Our field of interest concerns inverse scattering problems where the goal is the characterization of an unknown object (i.e., the determination of its location, its shape, or its constitutive physical parameters) from measurements of the scattered field which results from its interaction with a known interrogative (or incident) wave. Most of the time, in real practical applications, we are faced with *aspect-limited data configurations* when the unknown objects are embedded in intricate environments, e.g., inhomogeneous stratified media, so that they can be *seen* only from very restricted areas, i.e., the scattered field data can be collected only for a small number of sources and receivers which cannot be moved all around the object. Furthermore, in addition to their limited aspect, these data are often limited in frequency, since they are time-harmonic data collected at a single frequency or at a few discrete frequencies (frequency-diverse data).

This is the case of the two configurations considered herein where the unknown object to be retrieved is either immersed in a shallow sea water environment or embedded in the lower half-space of a two half-space fluid medium, which models a thick sediment layer in a deep sea environment, these situations being of increasing concern in underwater acoustics [GSWX98].

In order to achieve a trade-off between the expected resolution in the retrieved object and the penetration of the interrogating wave in the generally lossy embedding media, characterization of objects is usually done in the so-called *resonance domain* where the size of the object is of the same order of magnitude as the interrogating wavelength in the embedding, which prevents the introduction of high- or low-frequency approximations and leads us to describe the wave–object interaction through integral representations of the fields, which are suitable for studying scattering problems in the resonance domain. Hence, the wave–object interaction can be described by two coupled domains or contour integral equations, where, depending upon the nature of the object (penetrable or not), integration is made on the domain occupied by the object or on its boundary. Then, the inverse problem consists in retrieving, in a prescribed test domain assumed to contain the unknown object, a contrast function representative of its constitutive physical parameters, or its contour in the impenetrable case, from measurements of the scattered field through the inversion of the two aforementioned coupled integral equations.

It is well known that, as sketched above, inverse scattering problems are generally ill-posed, which means that uniqueness and stability of their solutions are not simultaneously guaranteed, and they must be regularized. As underlined in [Sab00], where a thorough investigation of the past and future of inverse problems is done, there is no universal method for solving ill-posed inverse problems. So, many solution methods, devoted to particular physical problems and whose efficiency is often restricted to these particular situations, have been developed these last few years, and a non-exhaustive review of such methods dedicated to the wave-field inversion of objects in stratified environments can be found in [LD96].

We revisit herein two solution methods tailored to the two configurations described above, which are precisely concerning objects in stratified environments. In these configurations, although the lack of information is partially alleviated for by considering frequency-diverse data, the inherent ill-posedness of the inverse problem is enhanced either by range filtering in the case of the waveguide configuration, which means that the high spatial frequency components of the scattered field are filtered out when the range of observation increases, or by the limited aspect of the data considered in the other case, which means that the data are collected only in the reflection mode and in a very restricted area. In order to overcome this ill-posedness, regularization is done by introducing, in the inversion algorithm, some a priori information on the object to be retrieved which consists in the smoothness of its contour or in its homogeneity.

Both of these methods are deterministic iterative techniques and try to solve a non-linear inverse problem. The first one, applied to the reconstruction of the contour of an impenetrable object, is the so-called *distributed source method*, which belongs to the class of complete family solution methods. It has been developed by Angell, Kleinman and Roach ([AKR87], [AKKR89]), and has been applied to the reconstruction of an object in free space [AJK97], as well as to the waveguide configuration considered here with an impenetrable sound-soft [RLAK97], and sound-hard [LL99] object, the mathematical setting concerning uniqueness and completeness in the latter configuration being found in [AKRL96].

The second method, devoted to the reconstruction of penetrable objects in aspect-limited data configurations, is the so-called *binary modified gradient method*, a specialization to the case of homogeneous objects of the modified gradient method. The latter, due to Kleinman and van den Berg ([KvdB92],[KvdB93]), belongs to the class of gradient-type methods, an insightful analysis of them being found in [KvdB97]. Several specializations of modified gradient method have been developed [KvdBDL97], whereas the binary one has been applied to various cases including ultrasonic and low frequency electromagnetic imaging of immersed objects [SDLK96], eddy current non-destructive evaluation in cylindrical [MDL98] and planar [MLD$^+$99] geometries, and microwave imaging of objects in free space from real data [DLK97].

We will not go, in this paper, into the details of the two aforementioned methods, the mathematical derivations and numerical details being left for the referenced contributions. Most of the theoretical analysis is borrowed from them and is, here, mostly descriptive, whereas new typical results of inversion are given in order to illustrate their behavior in 2-D configurations with cylindrical impenetrable or penetrable fluid objects, planar interfaces and homogeneous isotropic fluid media. These configurations are somewhat simple compared to those commonly encountered in real sea environments, but acquiring a precise knowledge of the environmental parameters would already constitute an inverse problem [CK94], and we rather focus on inverse obstacle scattering problems that are already hard to solve, even in the simplified configurations considered herein.

8.2   The Impenetrable Object in a Shallow-Water Open Waveguide

The goal is to retrieve the shape of an impenetrable, sound-hard or sound-soft object, described by its contour Γ, immersed in a shallow water waveguide, from measurements of the scattered pressure field which results from its interaction with a known incident wave. The pressure field is measured by means of one or two

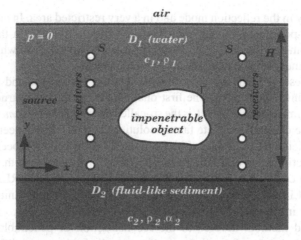

FIGURE 8.1. The impenetrable object in the open waveguide configuration.

vertical arrays of receivers and the incident field is produced by means of a single time-harmonic line source located in the near field of the object. The waveguide is a range-independent homogeneous, isotropic, lossless water layer D_1 of depth H, density ρ_1, and sound speed c_1, with a pressure release air–water interface and a penetrable fluidlike bottom D_2 (open waveguide) with density ρ_2, sound speed c_2, and attenuation α_2, which models a thick sediment layer (cf. Figure 8.1). Let us notice that an impenetrable hard bottom or a layered solid elastic one could be considered as well, because the modelization only involves the plane wave reflection coefficient at the water–bottom interface.

8.2.1 The Direct Model

By applying the Green theorem to the Helmholtz wave equation satisfied by the field in the waveguide and by accounting for the boundary conditions at its walls, we can obtain two coupled boundary integral equations. The first one, known as the *observation* or *data equation*, expresses the scattered pressure field p^s ($p^s(\mathbf{r}) = p(\mathbf{r}) - p^i(\mathbf{r})$, where p is the total field, i.e., the field in the presence of the object, and p^i is the incident field, i.e., the field in the absence of the object) observed on the array as a function of the total pressure field p and its normal derivative $\partial_n p$ ($\partial_n = \mathbf{n} \cdot \nabla$, $\mathbf{n}$ being the outgoing normal unit vector) on the boundary Γ of the object

$$p^s(\mathbf{r}) = \oint_\Gamma p(\mathbf{r}')\partial_n G(\mathbf{r}, \mathbf{r}')d\mathbf{r}' - \oint_\Gamma \partial_n p(\mathbf{r}')G(\mathbf{r}, \mathbf{r}')d\mathbf{r}', \qquad \mathbf{r} \in D_1, \qquad (8.1)$$

where $G(\mathbf{r}, \mathbf{r}')$ and $\partial_n G(\mathbf{r}, \mathbf{r}')$ are the Green function of the waveguide and its normal derivative, respectively. The Green function represents the field radiated by a line-source located at $\mathbf{r}'$ and observed at $\mathbf{r}$ in the absence of the object. It satisfies the Helmholtz wave equation in the waveguide and the wall boundary conditions.

Depending upon the nature of the bottom wall and upon the range of observation, several representations [Buc92] of this Green function are used; for an impenetrable bottom [RLAK97] a normal mode expansion is used when the distance between the source and the observation is large enough, a finite number of guided modes being considered in practice, whereas a hybrid ray-mode representation [Lu89] is used otherwise, a finite number of modes and rays being used then. When a penetrable bottom is considered, the Green function is obtained by means of a Fourier transform from its spectral representation in the (k_x, k_y) spatial frequency plane [LL99].

Let us notice that, in the configuration considered here, the object is illuminated by a line source located at r_e, so that the incident field p^i in the waveguide is given by

$$p^i(\mathbf{r}) = G(\mathbf{r}, \mathbf{r}_e), \qquad \mathbf{r} \in D_1, \tag{8.2}$$

The second equation, known as the *coupling* or *state equation*, links the total field on the boundary of the object to itself and to its normal derivative

$$\tfrac{1}{2}p(\mathbf{r}) = p^i(\mathbf{r}) + PV \oint_\Gamma p(\mathbf{r}')\partial_n G(\mathbf{r}, \mathbf{r}')d\mathbf{r}' - \oint_\Gamma \partial_n p(\mathbf{r}')G(\mathbf{r}, \mathbf{r}')d\mathbf{r}', \qquad \mathbf{r} \in \Gamma, \tag{8.3}$$

where PV stands for the principal value. In the case of a sound-soft object (henceforth denoted as the Dirichlet case), the field satisfies a Dirichlet boundary condition on Γ ($p(\mathbf{r}) = 0, \mathbf{r} \in \Gamma$), and these equations become:

$$p^s(\mathbf{r}) = -\oint_\Gamma \partial_n p(\mathbf{r}')G(\mathbf{r}, \mathbf{r}')d\mathbf{r}', \qquad \mathbf{r} \in D_1 \tag{8.4}$$

$$p^i(\mathbf{r}) = \oint_\Gamma \partial_n p(\mathbf{r}')G(\mathbf{r}, \mathbf{r}')d\mathbf{r}', \qquad \mathbf{r} \in \Gamma, \tag{8.5}$$

whereas for a sound-hard object (henceforth denoted as the Neumann case) the field satisfies a Neumann boundary condition on Γ ($\partial_n p(\mathbf{r}) = 0, \mathbf{r} \in \Gamma$), and they read:

$$p^s(\mathbf{r}) = \oint_\Gamma p(\mathbf{r}')\partial_n G(\mathbf{r}, \mathbf{r}')d\mathbf{r}', \qquad \mathbf{r} \in D_1 \tag{8.6}$$

$$\tfrac{1}{2}p(\mathbf{r}) = p^i(\mathbf{r}) + PV \oint_\Gamma p(\mathbf{r}')\partial_n G(\mathbf{r}, \mathbf{r}')d\mathbf{r}', \qquad \mathbf{r} \in \Gamma. \tag{8.7}$$

8.2.2   The Distributed Source Method

At the basis of the distributed source method is the appropriate definition of complete family of linearly independent radiating solutions of the Helmholtz wave equation in the waveguide. Here, these fundamental solutions are taken to be the Green functions of the waveguide $(G(\mathbf{r}, \mathbf{r}_m), m = 1, \ldots, \infty)$ whose sources are located at a countable dense set of points $(\mathbf{r}_m)$ on a closed curve Γ_{int} lying inside

the object cross-section. Of course, by definition, Green functions are radiating solutions of the Helmholtz wave equation in the waveguide. As for the completeness
of this set of solutions, it has been established for sound-soft objects (Dirichlet)
in the waveguide with an impenetrable bottom under restricting conditions on the
shape of the object, but it has not been established yet in the Neumann case.

In the numerical practice, the contour Γ is represented in polar coordinates
($\mathbf{r} = (r, \theta)$) by a trigonometric expansion $r = \gamma(\theta)$:

$$\gamma(\theta) = a_0 + \sum_{n=1}^{N} a_n \cos(n\theta) + \sum_{n=1}^{N-1} a_{N+n} \sin(n\theta), \qquad (8.8)$$

which implies that the object is *star-shaped* with respect to a given interior point.
This convenient representation constitutes a strong a priori information on the
object by reducing the class to which its contour is supposed to belong, but is not
imposed by the complete family inversion method, the latter only requiring that
the object is of smooth contour (at least C_2).

On the other hand the scattered field p^s is taken as a finite weighted sum of Green
functions with sources lying on a closed curve inside the object cross-section(cf.
Figure 8.2):

$$p^s(\mathbf{r}) = \sum_{m=1}^{M} c_m G(\mathbf{r}, \mathbf{r}_m^\gamma), \qquad \mathbf{r} \in S, \mathbf{r}_m^\gamma \in \Gamma_{int} \qquad (8.9)$$

and this curve moves during the iterative process, but it is kept homothetic to the
reconstructed contour, and at a close distance from it in terms of the wavelength
($r_m^\gamma(\theta) = \alpha\gamma(\theta), \alpha < 1$), if possible.

Now, the inverse problem can be reformulated as finding the real coefficients a_n
of the trigonometric expansion $\gamma(\theta)$ of Γ and the complex weights c_m of the Green
function expansion of the scattered field such that they minimize, in an appropriate
L_2 norm, a well-chosen cost functional F. This cost functional is a weighted sum
of two terms $F = f_1 + \sigma f_2$, σ being a weighting parameter. The first term

$$f_1 = \frac{\int_S |p^s(\mathbf{r}) - p^{\text{mes}}(\mathbf{r})|^2 d\mathbf{r}}{\int_S |p^{\text{mes}}(\mathbf{r})|^2 d\mathbf{r}}, \qquad \mathbf{r} \in S \qquad (8.10)$$

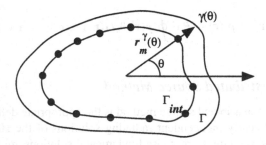

FIGURE 8.2. The closed curve Γ_{int}, lying inside the object cross-section and homothetic
to its contour Γ, is the support of the sources $\mathbf{r}_m^\gamma$.

is the residual on the observation equation normalized with respect to the data. It expresses the discrepancy between the scattered field p^s due to the retrieved object, computed by means of equation (8.9), and the data p^{mes} measured on the arrays S. The second term

$$f_2 = \frac{\int_0^{2\pi} |U(r, \theta)|^2 J_\Gamma(\theta)\, d\theta}{\int_0^{2\pi} |U^i(r, \theta)|^2 J_\Gamma(\theta)\, d\theta}, \quad (r, \theta) \in \Gamma, \quad U = \begin{cases} p, & \text{Dirichlet case,} \\ \partial_n p, & \text{Neumann case,} \end{cases} \tag{8.11}$$

measures how well the boundary condition (8.5) or (8.7) is satisfied on the contour of the retrieved object, the normalization being performed with respect to the incident field or its normal derivative. $J_\Gamma(\theta)$ is the Jacobian of the contour

$$J_\Gamma(\theta) = \sqrt{r^2 + \left(\frac{dr}{d\theta}\right)^2}, \tag{8.12}$$

which allows us to perform integrations on the *unit circle* rather than on the boundary Γ, this unit circle being independent of the unknown contour as the shape dependence is accounted for by the Jacobian $J_\Gamma(\theta)$.

In the numerical practice the integrals involved in (8.11) are computed by means of a trapezoidal rule; Q discrete test points are prescribed on Γ, and, for simplicity, Q is henceforth equated to the order M of the Green function expansion of the scattered field, so that the source locations r_m^γ on the homothetic contour Γ_{int} correspond point-to-point to the test points on Γ at prescribed θ_m.

The cost functional F is minimized iteratively by means of a Levenberg–Marquardt algorithm. Let us notice that the minimization of F is a highly nonlinear optimization problem, as the coefficients c_m depend upon the coefficients a_n via the location of the sources r_m^γ.

The data collected through several experiments, by varying the location of the source and/or of the receiver arrays, can be processed simultaneously by minimizing a global cost functional resulting from the summation of their respective costs. As for frequency-diverse data, they are processed with a *frequency-hopping* scheme, i.e., the different frequencies are treated sequentially, beginning with the lowest one, and the results obtained at a given frequency are used to initialize the inversion at the next frequency.

8.2.3 Initialization

As sketched above, the inversion scheme involves various tuning or weighting parameters as well as some variables that need to be appropriately set or initialized. The values used herein are inferred from [RLAK97], where an in-depth study of their influence is carried out in the Dirichlet case, and can be considered as *optimal* with respect to the trade-off between accuracy and computation time.

Hence, the optimal value of the homothetic parameter α has been found to be 0.75, whereas the weighting parameter σ is generally set to 1, except when noisy data are considered, where it can be chosen as close to, but greater than, 1 in order to favor the satisfaction of the boundary condition. Although the value of σ could be changed in the course of the iterative process, and a procedure such that an L-curve-type method [RL96] could be used to find an optimal value, this would be rather difficult to implement and computationally very costly.

As for the number M of sources in the Green function expansion of the scattered field or, equivalently, the number Q of test points on the contour Γ, it has to be chosen in order that both a good data fit is obtained on the receiver arrays when the exact contour is retrieved, and the boundary condition is closely satisfied on the entire contour Γ, the latter condition being the harder to satisfy. This leads us to take M such that the contour is sampled at a step lower than half the interrogating wavelength in the Dirichlet case, but approximately four times more sources have to be considered in the Neumann case [LL99].

In the absence of any other information, the initial contour is taken as a circle of radius R_0 (i.e., $a_0 = R_0$, $a_n = 0$, $n \geq 1$), which yields good results as far as its size is close to the one of the unknown object, but it can lead to unacceptable results otherwise. Finally, the amplitudes of the sources in the Green function expansion of the scattered field are initially set to 0 ($c_m = 0, m = 1, \ldots, M$). With a frequency-hopping scheme, all the sources are reset to 0 at each frequency, their number M being kept as constant, whereas the order N of the trigonometric expansion is increased at each frequency, since the resolution is expected to be better at higher frequencies. Hence the coefficients a_n obtained at the end of the iterative process corresponding to a given frequency are used as initial values for the first coefficients of the expansion at the next frequency, the remaining coefficients being set to 0.

8.2.4 Numerical Results

The method described above is applied to an open waveguide configuration, a water column of height $H = 100$ m, and of sound speed $c_1 = 1500$ m/s, and a fluidlike lossy penetrable bottom with sound speed $c_2 = 1650$ m/s, density $\rho_2 = 1.8$, and attenuation $\alpha_2 = 0.3$ dB/m/kHz. The operating frequency is either 30 Hz (wavelength in the water $\lambda_1 = 50$ m) or 100 Hz ($\lambda_1 = 15$ m). The object is a 20 m sided quasi-square centered in the middle of the waveguide, whose contour is described by a trigonometric expansion of order $N = 23$. The latter is searched as a trigonometric expansion of order $N = 4$ (which means that we look for eight real coefficients a_n) at 30 Hz and $N = 6$ at 100 Hz (starting from an initial circular boundary of radius $a_0 = 9$ m), whereas 18 Green functions are used in the expansion of the scattered field (which means that we look for 18 complex coefficients c_m).

The source is located on the left-hand side of the object in the middle of the waveguide ($y_e = 55$ m) and the data consist in the scattered pressure field measured, with a sampling step $\Delta y = 2.5$ m, by means of two vertical arrays S of 41 equally spaced receivers, which span the entire waveguide height on both sides of the object. The source and the receivers are located in the near field of the object ($x_e = -100$ m, $x_S = \pm 40$ m).

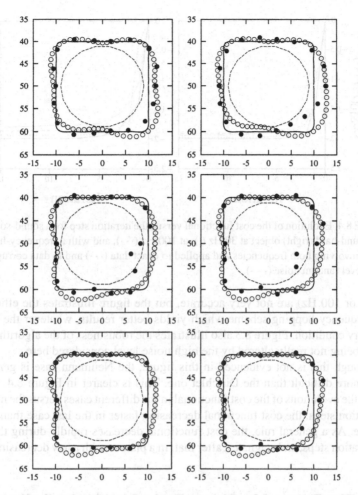

FIGURE 8.3. True (-), initial (- -), and retrieved (sound-soft: ●, sound-hard: ○) contours of an impenetrable cylindrical 20 m sided quasi-square object immersed in a horizontal sea channel (refer to Figure 8.1, at 30 Hz (top), at 100 Hz (middle), and with both frequencies (bottom), a frequency-hopping scheme being applied. The results are obtained from exact data (left) and from noisy data (right).

Synthetic data are used in the inversion. They are computed by means of the boundary integral equations (8.4)–(8.5) or (8.6)–(8.7) discretized using the Nyström method [Kre90] in order to avoid an inverse crime in the sense of [CK92].

Figure 8.3 shows the comparison between the results obtained after 300 iteration steps for the Neumann and Dirichlet cases, at 30 Hz, at 100 Hz, and with both frequencies, from exact data, and from noisy data. In the latter case, the real and imaginary parts of the scattered fields have been corrupted independently by an additive random noise whose maximum amplitude is 20% of the maximum amplitude of the signal at each frequency. The results obtained from a single-frequency dataset

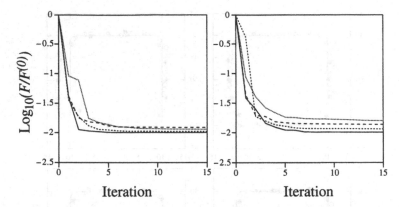

FIGURE 8.4. Evolution of the cost functional versus the iteration step for a sound-soft (left) and a sound-hard (right) object at 30 Hz ( ), at 100 Hz ($\cdots$), and with a frequency-hopping scheme involving both frequencies and applied to exact data ($\cdots$) and to data corrupted by a 20% level random noise (- - -).

(30 Hz or 100 Hz) are not very accurate, but the figure illustrates the efficiency of a frequency-hopping scheme which yields better results, whatever the object boundary condition. Figure 8.3 also illustrates the robustness of the algorithm, the results being not really altered by the high noise level considered here.

Although this is not evidenced in this figure, the Neumann case is generally much more difficult than the Dirichlet one. This is clearer in Figure 8.4, which depicts the evolutions of the cost functional in the different cases versus the number of iteration steps, the cost functional decreasing faster in the last case than in the first one. As a general rule, the cost functional decreases rapidly during the first few iteration steps, and stagnates, after that, in a plateau with a low decreasing rate.

8.3   The Penetrable Object Embedded in the Sediment

Let us now consider a penetrable object embedded in the lower half D_2 of a two half-space fluid medium, the source and the receivers, which operate at several discrete frequencies, being located in the upper half-space D_1 (cf. Figure 8.5). The density is now taken as constant throughout space, and the media are characterized by their complex propagation constants k_i ($k_i = \omega/c_i + j\alpha_i$) at angular frequency ω (the time-dependence $\exp(-j\omega t)$ is implied), whereas the object is characterized by a space-dependent contrast function $\chi(\mathbf{r}) = \left(k_\Omega^2(\mathbf{r}) - k_2^2\right)$ null outside the domain Ω.

8.3.1   The Direct Model

As in the previous section the modelization of the wave–object interaction is based upon two coupled integral equations, but domain integrals are now involved. Hence

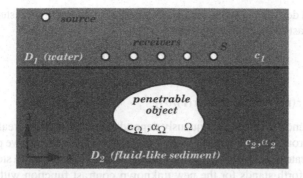

FIGURE 8.5. The aspect-limited data configuration with a penetrable object embedded in a fluidlike sediment, the source and the receivers being located in water.

the observation equation, which expresses the field observed on the receivers, now reads

$$p^s(\mathbf{r}) = \int_\Omega \chi(\mathbf{r}')p(\mathbf{r}')G_{12}(\mathbf{r}, \mathbf{r}')d\mathbf{r}', \qquad \mathbf{r} \in D_1, \qquad (8.13)$$

where integration is done in the domain Ω occupied by the object, and where G_{12} is the Green function with source in D_2 and observation in D_1. In the same way, the coupling equation, which expresses the total field inside the object, reads

$$p(\mathbf{r}) = p^i(\mathbf{r}) + \int_\Omega \chi(\mathbf{r}')p(\mathbf{r}')G_{22}(\mathbf{r}, \mathbf{r}')d\mathbf{r}', \qquad \mathbf{r} \in \Omega, \qquad (8.14)$$

where G_{22} is the Green function with both source and observation in D_2. These equations can be rewritten in a symbolic operator notation

$$p^s = G_S\chi p, \qquad p^i = [I - G_D\chi]\,p, \qquad (8.15)$$

where I is the identity operator and G_S and G_D are the integral operators as defined in (8.13) and (8.14), respectively.

8.3.2   The Binary Constraint

The inverse problem consists in retrieving the contrast function $\chi(\mathbf{r})$ ($\mathbf{r} \in D$), in a prescribed test domain D assumed to contain the object, from measurements of the scattered field p^s. As underlined before, this problem is ill-posed and the conditions for uniqueness of the solution are not yet established in the aspect-limited data configuration considered here. The problem is regularized by introducing the a priori information that the object is homogeneous and of known contrast, but of unknown shape and location, so that, after a suitable normalization, we look for a distribution in space of binary contrast values (1 inside the object and 0 elsewhere), i.e., for the characteristic function of the support of the contrast. However, the contrast is not directly sought in a binary form, as this would prevent the existence of functional derivatives needed in the modified gradient method. Instead, we look for the contrast in the following form

$$\chi(\mathbf{r}) = \chi_\Omega \psi_\theta(\tau(\mathbf{r})),$$

where

$$\psi_\theta(\tau(\mathbf{r})) = [1 + \exp(-\tau(\mathbf{r})/\theta)]^{-1}, \qquad \chi_\Omega = (k_\Omega^2 - k_2^2), \qquad (8.16)$$

that is, as a function which continuously depends upon an auxiliary real variable τ, ψ_θ running from 0 to 1 as τ varies from $-\infty$ to $+\infty$. The real positive parameter θ controls the rate of variation of ψ_θ: the lower θ is the closer is ψ_θ to a step function, and τ henceforth stands for the new unknown contrast function with respect to which the functional derivatives of interest are formally derived. Furthermore, after this transformation, the object needs no longer to be of known contrast, as the algorithm requires only an upper bound χ_Ω on the contrast to be retrieved and, although this has not yet been verified, we can also assume that the object does not need to be homogeneous anymore, in which case the algorithm is supposed to produce values of the contrast in between 0 and 1.

8.3.3 The Modified Gradient Method

In addition to its ill-posedness, the inverse problem at hand is also nonlinear, since in the observation equation, which links the scattered field to the contrast function χ, the total field inside the object shows up, and this total field also depends upon the contrast as indicated by the coupling equation. The modified gradient method handles this nonlinearity by simultaneously looking for the unknown contrast, more precisely here for τ, and for the unknown total fields within the object. They are sought iteratively with a gradient-type technique

$$\begin{aligned} p_q^{(n)}(\mathbf{r}) &= p_q^{(n-1)}(\mathbf{r}) + \alpha_q^{(n)} V_q^{(n)}(\mathbf{r}) \\ \tau^{(n)}(\mathbf{r}) &= \tau^{(n-1)}(\mathbf{r}) + \beta^{(n)} \xi^{(n)}(\mathbf{r}), \end{aligned} \qquad \mathbf{r} \in D, \qquad (8.17)$$

where superscript n stands for the iteration step and subscript q for the frequency, and where the coefficients $\alpha_q^{(n)}$ and $\beta^{(n)}$ are obtained by minimizing, by means of a Powell conjugate-gradient algorithm, a cost functional F which, here again, comprises two terms $F = f_1 + f_2$, i.e.,

$$f_1 = \sum_{q=1}^{N_q} |p_q^s - p_q^{\text{mes}}|_S^2 \times |p_q^{\text{mes}}|_S^{-2},$$

$$f_2 = \sum_{q=1}^{N_q} \|[I - G_{Dq}\chi_\Omega\psi_\theta(\tau)]p_q - p_q^i|_D^2 \times |p_q^i|_D^{-2}, \qquad (8.18)$$

with $\|\cdot\|_W$ being the norm associated to the inner product $\langle\cdot,\cdot\rangle_W$ in $L_2(W)$, $W = S$, or D, which represents the residuals of both the observation and the coupling equations, i.e., the discrepancies between the data (scattered field data p_q^{mes} and incident field data p_q^i) and the fields computed by means of the current solution, normalized with respect to the data. Let us notice that, when frequency-diverse data are considered, the cost functional is the sum of the cost functionals corresponding to the different frequencies.

As for V_q and ξ, although other choices can be made (cf. [KvdBDL97]), they are the update directions of a Polak–Ribière conjugate-gradient scheme, assuming that, respectively, τ or p_q does not change, i.e.,

$$V_q^{(n)} = g_q^{(n)} + \langle g_q^{(n)}, g_q^{(n)} - g_q^{(n-1)} \rangle_D \times \|g_q^{(n-1)}\|_D^{-2} V_q^{(n-1)},$$

$$\xi^{(n)} = h^{(n)} + \langle h^{(n)}, h^{(n)} - h^{(n-1)} \rangle_D \times \|h^{(n-1)}\|_D^{-2} \xi^{(n-1)}, \qquad (8.19)$$

with $g_q^{(n)}$ and $h^{(n)}$ being the opposites of the gradients of F with respect to p_q and τ, evaluated at the $(n-1)$th step (cf. [SDLK96]).

8.3.4   Cooling and Initialization

As stated above the contrast function $\psi_\theta(\tau)$ depends upon a real parameter τ which controls its rate of variation. This parameter can be tuned during the iterative process and this allows us to control its evolution by a so-called *cooling* procedure. It consists in decreasing the value of θ each time a plateau with a low decreasing rate of the cost functional is observed or after a given number of iterations, which allows us to push the values of the contrast in the test domain toward either 0 or 1. The cooling operation implies that all quantities of interest must be calculated again with the contrast corresponding to the new θ, and this can produce an instantaneous increase of the cost functional. However, on condition that a good cooling scheme is found (too large a decrease of θ would freeze the optimization process, whereas too small a decrease would be ineffective), this procedure is very effective in escaping from suboptimal solutions and, at the end of the iterative process the cost is generally at a much lower level than it would be without cooling, or with the standard modified gradient algorithm which does not take into account the binarity of the sought contrast, as evidenced in Figure 8.7.

As evidenced in the same figure, an instantaneous increase of the cost functional is also observed when the update directions are refreshed in the directions of the gradients, as is often necessary with conjugate-gradient schemes due to the loss of *orthogonality* of the successive update directions linked to numerical inaccuracies. However, contrary to cooling, although it is necessary, frequent refreshment often does not help to escape from local minima but rather illustrates the difficulties in achieving convergence of the optimization process.

In addition to a good cooling scheme, appropriate initialization of θ is also of importance for the success of the inversion. A good choice is to take its initial value so as to precondition the minimization of F by scaling the components $g_q^{(1)}$ and $h^{(1)}$ of its gradient versus p_q and τ, which are chosen as initial update directions $(V_q^{(1)} = g_q^{(1)}, \xi^{(1)} = h^{(1)})$, to the same order of magnitude.

Of course, initialization of the unknowns is also of importance. When lossless or weakly attenuating media are considered, the contrast and the fields can be initialized by means of a so-called *backpropagation* procedure. It consists in starting from an initial estimate $J_q^{(0)}$ of the contrast sources, whose constitutive relationship is $J_q(\mathbf{r}) = \chi_\Omega \psi_\theta(\tau(\mathbf{r})) p_q(\mathbf{r})$, within the test domain. This initial estimate is obtained by backpropagating the scattered field data from the measurement line S onto the test domain D:

$$J_q^{(0)}(\mathbf{r}) = \gamma G_{Sq}^* p_q^{\text{mes}}(\mathbf{r}), \qquad \mathbf{r} \in D, \tag{8.20}$$

where G_{Sq}^*, which acts from S onto D, is the operator adjoint to G_{Sq} such that $\left(G_{Sq} u, v\right)_S = \left(u, G_{Sq}^* v\right)_D$, u being a function in $L_2(D)$ and v a function in $L_2(S)$. The constant γ is determined by minimizing the quantity $\sum_q \| p_q^{\text{mes}} - \gamma G_{Sq} G_{Sq}^* p_q^{\text{mes}} \|_S^2$:

$$\gamma = \frac{\displaystyle\sum_{q=1}^{N_q} \langle p_q^{\text{mes}}, G_{Sq} G_{Sq}^* p_q^{\text{mes}} \rangle_S}{\displaystyle\sum_{q=1}^{N_q} \| G_{Sq} G_{Sq}^* p_q^{\text{mes}} \|_S^2}. \tag{8.21}$$

Then the initial estimate of the field $p_q^{(0)}$ is obtained by inserting $J_q^{(0)}$ in the coupling equation

$$p_q^{(0)}(\mathbf{r}) = p_q^i(\mathbf{r}) + G_{Dq} J_q^{(0)}(\mathbf{r}), \qquad \mathbf{r} \in D, \tag{8.22}$$

and the initial estimate of the contrast follows by minimizing the error in the contrast-source constitutive relationship

$$\psi_\theta(\tau^{(0)}(\mathbf{r})) = \sqrt{\frac{\displaystyle\sum_{q=1}^{N_q} \{\Re[J_q^{(0)}(\mathbf{r}) \overline{p}_q^{(0)}(\mathbf{r})] / |p_q^{(0)}(\mathbf{r})|\}^2}{\displaystyle\sum_{q=1}^{N_q} |p_q^{(0)}(\mathbf{r})|^2}} \tag{8.23}$$

where the overbar denotes the complex conjugate.

This procedure, which is very effective in lossless configurations, does not yield good results when strongly attenuating media are considered as backpropagation results in an enhancement of the effect of the noise. In that case, it is better to take as initial estimates of the fields the Born approximated ones ($p_q^{(0)}(\mathbf{r}) = p_q^i(\mathbf{r})$, $\mathbf{r} \in D$) and to take an initial contrast constant over the test domain ($\psi_\theta(\tau^{(0)}(\mathbf{r})) = \Xi$, $0 < \Xi < 1, \forall \mathbf{r} \in D$).

8.3.5 Numerical Results

The numerical study of the inverse problem is led from discrete counterparts of the above equations obtained by applying a Method of Moments with pulse-basis and point matching. This results in dividing the test domain in $N_x \times N_y = N_D$ elementary square pixels, the fields and the contrast being considered as constant over each pixel and their values at the center of the pixels being taken as the unknowns.

The data for the inverse problem are obtained by solving the associated direct problem, i.e., the discrete counterparts of the observation (8.3) and coupling (8.4) equations, χ being known, inverse crime being partly avoided by using a much finer discretization than for the inverse problem and by considering an object that does not exactly fit to the discretization mesh used in the latter. This problem is well-posed and can be efficiently solved by means of FFT-based iterative algorithms (cf. [LD91]) due to the convolutional/correlational forms of the equations.

We consider a configuration with media similar to those of the previous section except that density is constant throughout space ($\rho_i = 1, i = 1, 2, \Omega$). The upper half-space is made of lossless water with sound speed $c_1 = 1500$ m/s, and the lower half-space is made of a fluidlike lossy penetrable sediment with sound speed $c_2 = 1650$ m/s and attenuation $\alpha_2 = 0.3$ dB/m/kHz. The object is a lossless penetrable 20.5 m sided square whose center is at a depth of 20 m under the water–sediment interface, and whose sound speed is $c_\Omega = 1800$ m/s. It is illuminated by a plane wave under normal incidence at six different operating frequencies (30, 45, 60, 75, 90, and 100 Hz), the corresponding wavelengths in the water lying in between 15 m and 50 m. The fields are measured by means of an array of 64 receivers with a 2 m step, parallel to the water–sediment interface and located in water 2 m above this interface. The test domain is a 40 m sided square tangent to the water–sediment interface and partitioned into $N_D = 20 \times 20 = 400$ pixels. Test domain, object and receiver array are all centered on the $x = 0$ axis.

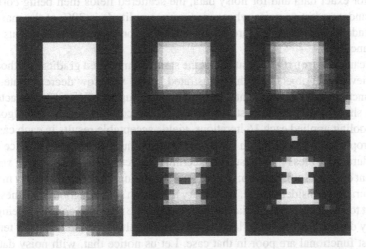

FIGURE 8.6. Gray-level maps of the contrast (0 corresponds to black and 1 corresponds to white) retrieved after 128 iterations in the case of the configuration depicted in Figure 8.5: the exact contrast (top-left) is compared to the contrasts retrieved with the standard modified gradient algorithm (bottom-left) and with the binary one (middle) initialized with a constant-contrast procedure (middle-top) or with backpropagation (middle-bottom). The robustness of the binary algorithm is illustrated in the right column, the data being corrupted by a 20% level random noise, for the constant-contrast initialization procedure (right-top) or for the backpropagation initialization procedure (right-bottom).

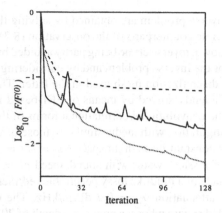

FIGURE 8.7. Evolution of the cost functional versus the iteration step for the standard modified gradient algorithm (solid line) and for the binary algorithm applied to exact data (dotted line) and to noisy data (dashed line), the procedure being initialized with a constant contrast. Spikes of the cost functional are due to refreshment of the update directions for the standard algorithm and to cooling for the binary one.

Figure 8.6 depicts the results obtained in the above configuration after 128 iteration steps with the standard modified gradient method and with the binary algorithm, and a backpropagation initialization procedure or a constant initial contrast, for exact data and for noisy data, the scattered fields then being corrupted by a random additive noise whose maximum amplitude is 20% of the maximum amplitude of the signal. Figure 8.7 displays the corresponding evolutions of the cost functional.

We can infer from these results that the standard modified gradient method fails in retrieving the object, which is illustrated by the very low decrease rate of the cost functional and the frequent refreshments of the update directions characterized by the sharp spikes of the cost functional. On the contrary, the binary algorithm, with cooling applied each 16 iterations, yields acceptable results in each case: the backpropagation initialization scheme leads to a much faster convergence of the procedure but to less good results concerning the retrieved contrast than with the constant contrast initialization, which is not surprising in the weakly lossy medium configuration considered here. The binary algorithm shows a good robustness with respect to noise with both initialization schemes, the retrieved contrasts being only slightly distorted with a 20% level random noise, although the results in terms of the cost functional are poor in that case. Let us notice that, with noisy data, the latter may not be a good measure of convergence.

8.4   Conclusion

As underlined in the Introduction, the authors neither intended to exhaustively review the inversion methods that can be applied to the two aforementioned configurations, nor to consider the various intricate configurations that can be encountered in real sea environments. Rather, they intended to show the behavior of two methods, which they have thoroughly investigated elsewhere in other situations, in voluntarily simplified configurations in order to illustrate the intrinsic difficulty of inverse obstacle scattering problems in underwater acoustics.

Of course, other inversion methods could be used for each configuration, and particularly each one of the two methods applied herein to one particular configuration could be applied to the other. Hence, applying the distributed source method to the retrieval of the contour of a penetrable fluidlike obstacle seems to be feasible, although there is no theoretical justification, by considering two sets of equivalent sources, i.e., one inside the object domain to model the outer field, and the other outside in order to model the inner field, the continuity of the pressure being then the boundary condition enforced on the object contour. A similar approach for a direct modeling of a simpler electromagnetic case can be found in [LM90]. The fact that the object is located in the waveguide or totally embedded in the sediment does not really matter, as the differences lie in the Green functions to be considered. As for an object partially buried in the sediment, it cannot be sought with the distributed source method without theoretical difficulties linked to the wavefield representation, but it may be amenable to numerical calculations. To conclude with the distributed source method, as sketched above, the main drawbacks remain the requirements in terms of smoothness, star-shapedness, and knowledge of an interior point. Such requirements are avoided if the retrieval of the contour of a penetrable obstacle is tackled, for example, by the controlled evolution of a level set, an other nonlinearized iterative technique investigated by one of the authors in a simpler electromagnetic case [LLS98].

Although for an impenetrable object the only meaningful information that can be inferred from the measured fields comes from its boundary, the binary modified gradient algorithm can be applied, without major difficulties, to the retrieval of the shape of such an object as is exemplified by two of the authors in a low-frequency electromagnetic case [SDLK96]. In that case the upper bound on the contrast χ_Ω has to be set through physical and numerical considerations as stated in [KvdB94], where the optimal upper bound was found to be such that the penetration depth of the wavefield in the corresponding object is no smaller than three times the side of the pixels discretizing the test domain. For the same reasons as in the previous paragraph, the binary modified gradient method can be applied to the waveguide problem, and, unlike the distributed source method, it can easily handle the case of an object partially buried in the sediment. However, density variations cannot be considered without major difficulties, even for an homogeneous object, as this would lead us to look for additional field unknowns on the boundary of the obstacle, and thus to a much more involved computational burden. Besides, even in simple configurations, requirement in terms of computation time is one of the major drawbacks of the modified gradient method. Close to the latter and to its total-variation enhanced version [vdBK95], but much less time-demanding, are the contrast source inversion method [vdBK97] and its total-variation enhanced version [vdBvBA99], where the contrast sources J_q and the contrast itself are iteratively sought by alternate updating in contradistinction to the modified gradient method where both contrast and field are updated simultaneous.

As a general rule, the computation time is always a problem with nonlinearized iterative methods. However, their performances can be greatly enhanced, in terms of computation time as well as in terms of quality of the reconstruction, if a good initialization scheme is found; initialization can generally be done through faster approximated (e.g., linearized) methods, on condition that they are adapted to the configuration at hand. But, in the aspect-limited data configurations considered here, the key of the success of the inversion remains the a priori information on the solution we look for and how it can be introduced in the inversion algorithm.

Further challenging problems remain to be addressed, which result from a more realistic modeling of the sea environment or of the object. Hence a depth-varying water column could be considered at the prize of a much more involved computation of the Green function, whereas a range-varying one does not seem to be of interest in view of the ranges considered herein. As for the sea bottom, a layered solid (visco- or poro-)elastic one could be modelized as far as the waveguide problem is addressed, but would be much less tractable if the object is embedded inside. The case of a nonplanar (i.e., rough) sea surface and water–bottom interface seems to be out of reach. Finally, the first case of study would certainly be a three-dimensional object in the waveguide or in the bottom. A good example of such an object is a body of revolution, which has been successfully investigated in free space with the distributed source method [AJK97]; inversion then could appropriately be initialized through the intersecting canonical domain approximation [GSWX98].

8.5   References

[AJK97]   T.S. Angell, J. Jiang, and R.E. Kleinman. A distributed source method for inverse acoustic scattering. *Inverse Problems*, 13:531–546, 1997.

[AKKR89]   T.S. Angell, R.E. Kleinman, B.Kok, and G.F. Roach. A constructive method for identification of an impenetrable scatterer. *Wave Motion*, 11:185–200, 1989.

[AKR87]   T.S. Angell, R.E. Kleinman, and G.F. Roach. An inverse transmission problem for the Helmholtz equation. *Inverse Problems*, 3:149–180, 1987.

[AKRL96]   T.S. Angell, R.E. Kleinman, C. Rozier, and D. Lesselier. Uniqueness and complete families for an acoustic waveguide problem. Technical Report 96-4, Center for the Mathematics of Waves, University of Delaware, Newark, 1996.

[Buc92]   M.J. Buckingham. Ocean-acoustics propagation models. *J. of Acoust.*, **5**:223–287, 1992.

[CK92]   D. Colton and R. Kress. *Inverse Acoustic and Electromagnetic Scattering Theory.* Springer-Verlag, New York, 1992.

[CK94]   M.D. Collins and W.A. Kuperman. Inverse problems in ocean acoustics. *Inverse Problems*, **10**:1023–1040, 1994.

[DLK97]   B. Duchêne, D. Lesselier, and R.E. Kleinman. Inversion of the 1996 Ipswich data using binary specializations of modified gradient methods. *Antennas Propagation Mag.*, **39**:9–12, 1997.

[GSWX98]   R.P. Gilbert, T. Scotti, A. Wirgin, and Y.S. Xu. The unidentified object problem in a shallow ocean. *J. Acoust. Soc. Am.*, **103**:1320–1328, 1998.

[Kre90]   R. Kress. Numerical solution of boundary integral equations in the time-harmonic electromagnetic scattering. *Electromagnetics*, **10**:1–20, 1990.

[KvdB92]   R.E. Kleinman and P.M. van den Berg. A modified gradient method for two-dimensional problems in tomography. *J. Comput. Appl. Math.*, **42**:17–35, 1992.

[KvdB93]   R.E. Kleinman and P.M. van den Berg. An extended range modified gradient technique for profile inversion. *Radio Science*, **28**:877–884, 1993.

[KvdB94]   R.E. Kleinman and P.M. van den Berg. Two-dimensional location and shape reconstruction. *Radio Science*, **29**:1157–1169, 1994.

[KvdB97]   R.E. Kleinman and P.M. van den Berg. Gradient methods in inverse acoustic and electromagnetic scattering. In L.T. Biegler, T.F. Coleman, A.R. Conn, and F.N. Santosa, (eds.), *Large-Scale Optimization with Applications*, pp. 173–194. Springer-Verlag, Berlin, 1997.

[KvdBDL97] R.E. Kleinman, P.M. van den Berg, B. Duchêne, and D. Lesselier. Location and reconstruction of objects using a modified gradient approach. In G. Chavent and P.C. Sabatier, (eds.), *Inverse Problems of Wave Propagation and Diffraction*, pp. 143–158. Springer-Verlag, Berlin, 1997.

[LD91]   D. Lesselier and B. Duchêne. Buried two-dimensional penetrable objects illuminated by line sources: FFT-based iterative computations of the anomalous field. In T.K. Sarkar (ed.), *Application of Conjugate Gradient Methods to Electromagnetics and Signal Analysis*, pp. 400–438. Elsevier, New York, 1991.

[LD96]   D. Lesselier and B. Duchêne. Wavefield inversion of objects in stratified environments. From backpropagation schemes to full solutions. In W.R. Stone, (ed.), *Review of Radio Science 1993–1996*, pp. 235–268. Oxford University Press, Oxford, 1996.

[LL99]   M. Lambert and D. Lesselier. Distributed source method for retrieval of the cross-sectional contour of an impenetrable cylindrical obstacle immersed in a shallow water waveguide. To appear in *ACUSTICA—Acta Acustica*, 86 (4): 45–24. 2000.

[LLS98]   A. Litman, D. Lesselier, and F. Santosa. Reconstruction of a two-dimensional binary obstacle by controlled evolution of a level set. *Inverse Problems*, 14:685–706, 1998.

[LM90]   Y. Leviatan and Y. Meyouhas. Analysis of electromagnetic scattering from buried cylinders using a multifilament current model. *Radio Science*, 25:1231–1244, 1990.

[Lu89]   I.-T. Lu. Analysis of acoustic wave scattering by scatterers in layered media using the hybrid ray-mode (boundary integral equation) method. *J. Acoust. Soc. Am.*, 86:1136–1142, 1989.

[MDL98]   V. Monebhurrun, B. Duchêne, and D. Lesselier. Three-dimensional inversion of eddy current data for nondestructive evaluation of steam generator tubes. *Inverse Problems*, 14:707–724, 1998.

[MLD+99]   V. Monebhurrun, D. Lesselier, B. Duchêne, A. Ruosi, M. Valentino, G. Pepe, and G. Peluso. Eddy current nondestructive evaluation using SQUIDs. In D. Lesselier and A. Razek (eds.), *Electromagnetic Non-Destructive Evaluation (III)*, pp. 171–181. IOS Press, Amsterdam, 1999.

[RL96]   C. Rozier and D. Lesselier. Inversion of a cylindrical vibrating body in shallow water from aspect-limited data using filtered SVD and the L-curve. *ACUSTICA—Acta Acustica*, 82:717–728, 1996.

[RLAK97]   C. Rozier, D. Lesselier, T.S. Angell, and R.E. Kleinman. Shape retrieval of a cylindrical obstacle immersed in shallow water from single-frequency farfields using a complete family method. *Inverse Problems*, 13:487–508, 1997.

[Sab00]   P.C. Sabatier. Past and future of inverse problems. *J. Math. Phys.*, 2000, to appear.

[SDLK96]   L. Souriau, B. Duchêne, D. Lesselier, and R.E. Kleinman. A modified gradient approach to inverse scattering for binary objects in stratified media. *Inverse Problems*, 12:463–481, 1996.

[vdBK95]   P.M. van den Berg and R.E. Kleinman. A total variation enhanced modified gradient algorithm for profile reconstruction. *Inverse Problems*, 11:L5–10, 1995.

[vdBK97]   P.M. van den Berg and R.E. Kleinman. A contrast source inversion method. *Inverse Problems*, **13**:1607–1620, 1997.

[vdBvBA99] P.M. van den Berg, A.L. van Broekhoven, and A. Abubakar. Extended contrast source inversion. *Inverse Problems*, **15**:1325–1344, 1999.

9

Inverse Boundary-Value Problem for Ocean Acoustics Using Point Sources

Musaru Ikehata
George N. Makrakis
Gen Nakamura

ABSTRACT We formulate and prove uniqueness for the reconstruction of a compact inhomogeneity of the sound speed embedded in a uniform ocean acoustic channel of finite depth. As input data, we use the Green function on a vertical cylinder enclosing the inhomogeneity, at constant frequency.

9.1 Introduction

Let $\mathbb{R}_H^3 := \{\mathbf{x} = (x, y, z): 0 < z < H\}$ be a finite depth ocean with flat bottom $z = H$ and surface $z = 0$. Consider the idealized observation data obtained from two arrays of sonars on a wire rope hung from ships moving around in such a way that the trajectory of each wire rope forms the lateral surface of a cylinder $\Gamma := \{\mathbf{x} = (\mathbf{x}', z) = (x, y, z): |\mathbf{x}'| = R > 0, 0 < z < H\}$. The sonars of the first ship emit time-harmonic sound waves and the sonars of the second ship receive the sound waves. Then, our inverse problem is to identify the refraction index $n(\mathbf{x})$ of the ocean from the above described data. We will mathematically idealize this problem and we will give a more precise mathematical formulation. For this purpose, we make the following three assumptions:

(A-1) The refraction index $n \in L^\infty(\mathbb{R}_H^3)$ is constant, say one, outside the cylinder $B(R) := \{\mathbf{x} = (\mathbf{x}', z): |\mathbf{x}'| < R, 0 < z < H\}$.

(A-2) $k_0 > 0$, $k_0 \neq (2\ell + 1)\pi/2H$ ($\ell = 0, 1, 2, \ldots$).

(A-3) Zero is not an eigenvalue of the following boundary value problem:

$$\begin{cases} (\Delta + k_0^2 n^2(\mathbf{x}))u = 0 & \text{in } B(R) \\ \frac{\partial u}{\partial z}|_{z=H} = u|_{z=0} = 0 & \text{(referred to by b.c. from now on),} \\ u|_\Gamma = 0. \end{cases}$$

Now let $\phi_\ell(z) = \sin\{k_0^2(1 - a_\ell^2)^{1/2}z\}$ with $a_\ell = (1 - (2\ell+1)^2\pi^2/4k_0^2H^2)^{1/2}$ for $\ell = 0, 1, 2, \ldots$. Also, let $G(\mathbf{x}, \boldsymbol{\xi})$ be the outgoing Green function

$$\begin{cases} (\Delta + k_0^2 n^2(\mathbf{x}))G(\mathbf{x}, \boldsymbol{\xi}) + \delta(\mathbf{x} - \boldsymbol{\xi}) = 0 & \text{in } \mathbb{R}_H^3 \\ \frac{\partial G}{\partial z}\big|_{z=H} = G\big|_{z=0} = 0, \\ \lim_{|\mathbf{x}'|\to\infty} |\mathbf{x}'|^{1/2}(\frac{\partial G_\ell}{\partial|\mathbf{x}'|} - ik_0 a_\ell G_\ell) = 0 & (\ell = 0, 1, 2, \ldots) \end{cases}$$

where

$$G_\ell = G_\ell(\mathbf{x}, \boldsymbol{\xi}) = \frac{2}{H}\int_0^H G(\mathbf{x}', z, \boldsymbol{\xi})\phi_\ell(z)\,dz.$$

Here, the last condition is the so-called radiation condition and it will be referred to by r.c. from now on.

Remarks. (i) About the existence of the outgoing Green function, there are several recent results by Croc and Dermenjian [CD1], [CD2], [CD3] in the two-dimensional case for the operator $L_c\bullet := \mathrm{div}(c^2\mathrm{grad}\,\bullet) + k_0^2\bullet$. The relation

$$c^{-1}L_c(c^{-1}\bullet) = (\Delta - q)\bullet, \qquad q = c^{-1}\Delta(c) - k_0^2 c^{-2},$$

enables us to use their results for the Helmholtz equation of ocean acoustics. In the three-dimensional case, Xu [X] and Gilbert and [GX] constructed the outgoing Green function for $n = 1$ and $n = n(z)$ outside a sound-soft obstacle $\mathbb{R}_H^3$, respectively. None of their results can be used for our problem. Proving the existence of the outgoing Green function is the most tedious part of our work.

(ii) The Green function $G(\mathbf{x}, \boldsymbol{\xi})$ is just the Schwartz kernel of the Green operator $G: H_\delta^{-1}(\mathbb{R}_H^3) \longrightarrow \dot{H}_{-\delta}^1(\mathbb{R}_H^3)$ $(\frac{1}{2} < \delta < 1)$, defined as follows. For any $f \in H_\delta^{-1}(\mathbb{R}_H^3)$, $u := Gf \in \dot{H}_{-\delta}^1(\mathbb{R}_H^3)$ is defined as the solution to

$$\begin{cases} (\Delta + k_0^2 n^2(\mathbf{x}))u + f = 0 & \text{in } \mathbb{R}_H^3, \\ \text{b.c} \quad \text{and} \quad \text{r.c.,} \end{cases}$$

where $\dot{H}_\sigma^m(\mathbb{R}_H^3) := \{u: \langle\mathbf{x}\rangle^\sigma u \in H^m(\mathbb{R}_H^3), u|_{z=0}\}$ is our basic weighted Sobolev space, $H_{-\sigma}^{-m}(\mathbb{R}_H^3) := (\dot{H}_\sigma^m(\mathbb{R}_H^3))^*$ $(m \in \mathbb{N}\cup\{0\}, \sigma \in \mathbb{R})$ is the dual space, $H^m(\mathbb{R}_H^3)$ being the usual Sobolev space, and $\langle\mathbf{x}\rangle := \sqrt{1 + |\mathbf{x}|^2}$.

Note also that here the radiation condition abbreviated by r.c., has the meaning

$$\lim_{|\mathbf{x}'|\to\infty} |\mathbf{x}'|^{1/2}\left|\frac{\partial u_\ell}{\partial|\mathbf{x}'|}(\mathbf{x}') - ik_0 u_\ell(\mathbf{x}')\right| = 0,$$

where

$$u_\ell(\mathbf{x}') = \frac{2}{H}\int_0^H u(\mathbf{x}', z)\overline{\phi_\ell(z)}\,dz \qquad (\ell = 0, 1, \ldots).$$

The existence of the solution u is proven by using the standard argument (cf. [C]) of the limiting absorption principle and the following uniqueness lemma essentially given in [X]:

Lemma 9.1. If $u \in H^2_{loc}(\mathbb{R}^3_H) := \{u \in \mathcal{D}'(\mathbb{R}^3_H): \phi u \in H^2(\mathbb{R}^3_H) (\phi \in C^\infty_0(\overline{\mathbb{R}^3_H}))\}$ with $C^\infty_0(\overline{\mathbb{R}^3_H}) := \{\phi \in C^\infty(\mathbb{R}^3): supp \, \phi \cap \overline{\mathbb{R}^3_H}$ is compact$\}$: satisfies

$$\begin{cases} (\Delta + k_0^2 n^2(\mathbf{x}))u = 0 & in \, \mathbb{R}^3_H \\ b.c. \quad and \quad r.c., \end{cases}$$

then $u = 0$ in $\mathbb{R}^3_H$. Here $\mathcal{D}'(\mathbb{R}^3_H)$ is the space of distributions in $\mathbb{R}^3_H$.

(iii) The outgoing r.c. for the Green function follows from that of Gf for any $f \in H^{-1}_\delta(\mathbb{R}^3_H)$ by using the following fact (cf. Theorem 8.1 in [M]).

Let $0 < \varepsilon \ll 1$. For any ξ whose distance to the boundary of $\mathbb{R}^3_H$ is greater than 2ε,

$$G(\cdot, \xi) = \alpha_\varepsilon(\cdot - \xi)E(\cdot, \xi) + GL((1 - \alpha(\cdot - \xi))E(\cdot, \xi),$$

where $\alpha_\varepsilon(\mathbf{x}) = \alpha(\varepsilon^{-1}r)$ with $r := |\mathbf{x}|$, $\alpha \in C^\infty_0([0, \infty))$, $\alpha(r) = 1$ $(0 \le r \le \frac{1}{2})$, 1 $(r \ge 1)$, $L := \Delta + k_0^2 n^2(\mathbf{x})$, $E(\mathbf{x}, \xi)$ is given by

$$E(\mathbf{x}, \xi) := E_0(\mathbf{x}, \xi) + R(\mathbf{x}, \xi),$$

where $E_0(\mathbf{x}, \xi)$ is the Green function for $L_0 := \Delta + k_0^2$ satisfying, as a function of $\mathbf{x}$, both the b.c. and r.c. given in [X], and $R(\mathbf{x}, \xi) \in H^2(\mathbb{R}^3_H)$ is the solution to

$$\begin{cases} LR(\cdot, \xi) = -k_0^2(n^2 - 1)E(\cdot, \xi) & in \, \mathbb{R}^3_H \\ b.c. \quad and \quad r.c., \end{cases}$$

Now, we consider the following inverse problem.

Inverse problem. Identify $n(\mathbf{x})$ from $G(\mathbf{x}, \xi)$ for $\mathbf{x}, \xi \in \Gamma$ or in some parts of Γ $(\mathbf{x} \neq \xi)$.

Then, we have the following uniqueness results for the inverse problem.

Theorem 9.2. Assume (A-1) to (A-3). Then, the identification of $n(\mathbf{x})$ from $G(\mathbf{x}, \xi)$ ($\mathbf{x}, \xi \in \Gamma, \mathbf{x} \neq \xi$) is unique. Moreover, if $n(\mathbf{x})$ is one outside a cylinder $B(R')$ with $0 < R' < R$, then for any open subsets Σ_1, Σ_2 of Γ, the identification of $n(\mathbf{x})$ from $G(\mathbf{x}, \xi)$ ($\xi \in \Sigma_1, \mathbf{x} \in \Sigma_2, \mathbf{x} \neq \xi$) is unique. Here Σ_1, Σ_2 can be very small and equal.

Remarks. (i) The assumption $n(\mathbf{x}) = 1$ outside $B(R')$ for some $0 < R' < R$ for $n(\mathbf{x}) \in L^\infty(\mathbb{R}^3_H)$ will be referred as (A-1a) from now on.

(ii) $\mathbf{x}$ and ξ denote the positions of the receiver and emitter of the sound wave, respectively. We have idealized that the array of sonars are densely distributed along the wire ropes and the number of the observations is quite enough to consider that we can observe the sound at every point $\mathbf{x}$ in Γ (or Σ_2) emitted from every point ξ in Γ (or Σ_1). We have also idealized that the conditions used to derive our acoustic-wave equation does not change during the successive observations done one after another by moving the ships.

For the reconstruction of $n(\mathbf{x})$ we introduce two new assumptions (A-1b) and (A-4):

(A-1b) $\operatorname{supp}(n^2(\mathbf{x}) - 1) \subset B(R')$ for some $0 < R' < R$.

(A-4) Let $\Omega \subset B(R')$ be an open set with a smooth boundary $\partial\Omega$ such that $\overline{\Omega} \subset B(R')$, $\operatorname{supp}(n^2 - 1) \subset \Omega$. Then, zero is not an eigenvalue of the boundary value problems

$$\begin{cases} (\Delta + k_0^2 n^2(\mathbf{x}))u = 0 & \text{in } \Omega, \\ u|_{\partial\Omega} = 0, \end{cases}$$

and

$$\begin{cases} (\Delta + k_0^2)u = 0 & \text{in } B(R), \\ \text{b.c.,} \\ u|_\Gamma = 0. \end{cases}$$

Then, we have the following theorem about the reconstruction of $n(\mathbf{x})$:

Theorem 9.3. *Assume* (A-1b) *and* (A-2), (A-3) *and* (A-4). *Then, we can reconstruct $n(\mathbf{x})$ from $G(\mathbf{x}, \boldsymbol{\xi})$ ($\boldsymbol{\xi} \in \Sigma_1, \mathbf{x} \in \Sigma_2, \mathbf{x} \neq \boldsymbol{\xi}$).*

Remark. Xu [X] and Gilbert and Xu [GX] considered the inverse scattering by a soft obstacle using the far-field pattern of the propagating mode. However, there are several difficulties in proving the uniqueness and reconstruction, because they have to neglect the evanescent mode. So instead of using the far-field pattern of the propagating mode, we think it is better to use a near-field data such as point source data.

In the sequel we will give only the outline of the proofs of Theorems 9.2 and 9.3. The details of the proofs of the theorems and several facts used in this paper will be given elsewhere.

9.2   Outline of the Proof of Theorem 9.2

We reduce the proof of Theorem 9.2, to the following uniqueness result:

Theorem 9.4 (cf. [IMN]). *Let*

$$\Lambda : \dot{H}^{1/2}(\Gamma) \longrightarrow H^{-1/2}(\Gamma)$$

be the Dirichlet to Neumann map, formally defined by

$$\Lambda f := \left.\frac{\partial u(f)}{\partial \nu}\right|_\Gamma \in H^{-1/2}(\Gamma) := (\dot{H}^{1/2}(\Gamma))^*$$

for any

$$f \in \dot{H}^{1/2}(\Gamma) := \{\phi|_\Gamma : \phi \in H^1(\mathbb{R}_H^3), \ \phi|_{z=0} = 0, \ \text{supp } \phi \text{ is relatively compact}\},$$

where $(\dot{H}^{1/2}(\Gamma))^*$ denotes the dual space of $\dot{H}^{1/2}(\Gamma)$, ν is the outer unit normal vector of Γ, and $u = u(f) \in H^1(B(R))$ is the solution to

$$\begin{cases} \Delta + k_0^2 n^2(\mathbf{x}))u = 0 & \text{in } B(R), \\ b.c., \\ u|_\Gamma = f \in \dot{H}^{1/2}(\Gamma). \end{cases}$$

Then, the identification of $n(\mathbf{x})$ from Λ is unique.

To start our reduction argument, we remark that, by the analytic continuation, we know $G(\mathbf{x}, \boldsymbol{\xi})$ $(\mathbf{x}, \boldsymbol{\xi} \in \Gamma, \mathbf{x} \neq \boldsymbol{\xi})$ even for the case $G(\mathbf{x}, \boldsymbol{\xi})$ is only given for $\boldsymbol{\xi} \in \Sigma_1, \mathbf{x} \in \Sigma_2, \mathbf{x} \neq \boldsymbol{\xi}$ under the assumption (A-1').

Now, for any $f \in H^{-1/2}(\Gamma)$, define $T_f \in H^{-1}(\mathbb{R}_H^3) := (\dot{H}_0^1(\mathbb{R}_H^3))^*$ by $\langle T_f, \phi \rangle = \langle f, \phi|_\Gamma \rangle$ for any $\phi \in \dot{H}_0^1(\mathbb{R}_H^3)$. Also, for any $f \in H^{-1/2}(\Gamma)$, define S by $Sf := u(f)|_\Gamma$, where $u = u(f) \in H^1(\mathbb{R}_H^3)$ is the solution to

$$\begin{cases} (\Delta + k_0^2 n^2(\mathbf{x}))u = -T_f & \text{in } \mathbb{R}_H^3, \\ b.c., \\ u|_\Gamma = 0, \\ r.c. \end{cases}$$

Note that Sf can be formally given by

$$Sf(\mathbf{x}) = \int_\Gamma G(\mathbf{x}, \boldsymbol{\xi}) f(\boldsymbol{\xi}) \, d\sigma_{\boldsymbol{\xi}} \qquad (\mathbf{x} \in \Gamma)$$

because $G(\mathbf{x}, \boldsymbol{\xi})$ is the Schwarz kernel of the outgoing Green function. Here $d\sigma_{\boldsymbol{\xi}}$ is the surface measure of Γ. Hence, S is the trace of the single layer potential operator. Moreover, define the exterior Dirichlet to Neumann map $\Lambda^e : \dot{H}^{1/2}(\Gamma) \longrightarrow H^{-1/2}(\Gamma)$ formally by

$$\Lambda^e f := - \left. \frac{\partial u(f)}{\partial \nu} \right|_\Gamma \in H^{-1/2}(\Gamma) \qquad \text{for any } f \in \dot{H}^{1/2}(\Gamma),$$

where $u = u(f) \in H^1(\mathbb{R}_H^3 \backslash \overline{B(R)})$ is the solution to

$$\begin{cases} (\Delta + k_0^2)u = 0 & \text{in } \mathbb{R}_H^3 \backslash \overline{B(R)}, \\ b.c., \\ u|_\Gamma = f, \\ r.c. \end{cases}$$

Remark. The existence of the solution u to define the exterior Dirichlet to Neumann map can be proven likewise to that of the solution u to define the outgoing Green operator G.

Then, we have the following key lemma from which we immediately have Theorem 9.2:

Lemma 9.5. (i) $\Lambda - \Lambda^e : \dot{H}^{1/2}(\Gamma) \longrightarrow H^{-1/2}(\Gamma)$ *is injective.*

(ii) $(\Lambda - \Lambda^e)S = I$ *(identity operator).*

Therefore, Λ is given by $\Lambda = \Lambda^e - S^{-1}$ and we know Λ from $G(\mathbf{x}, \boldsymbol{\xi})$ $(\mathbf{x}, \boldsymbol{\xi} \in \Gamma, \mathbf{x} \neq \boldsymbol{\xi})$ by using the definition of S.

9.3   Outline of the Proof of Theorem 9.3

Define the Dirichlet to Neumann map

$$\Lambda^- : H^{1/2}(\partial\Omega) \longrightarrow H^{-1/2}(\partial\Omega)$$

formally by

$$\Lambda^- f := \left.\frac{\partial u(f)}{\partial \nu}\right|_{\partial\Omega} \qquad (f \in H^{1/2}(\partial\Omega))$$

where ν is the outer (with respect to Ω) unit normal of $\partial\Omega$ and $u_- = u_-(f) \in H^1(\Omega)$ is the solution to

$$\begin{cases} (\Delta + k_0^2 n^2(\mathbf{x}))u_- = 0 & \text{in } \Omega, \\ u_-|_{\partial\Omega} = f. \end{cases}$$

Then, we can prove Theorem 9.3 by reducing it to the following theorem:

Theorem 9.6 (Nachman [N]).   *There is a procedure to reconstruct $n(\mathbf{x})$ from Λ^-.*

Remark. For the convenience for the readers, the Nachman reconstruction will be given later after finishing the reduction argument.

The reduction consists of three lemmas. This type of reduction argument was given in [I] to determine the Dirichlet to Neumann map for the interior subregion from that of the whole region if we know the coefficient of the equation between the two regions. To begin, define the Dirichlet to Neumann map $\Lambda^+ : H^{1/2}(\partial\Omega) \longrightarrow H^{-1/2}(\partial\Omega)$ formally by

$$\Lambda^+ f := -\left.\frac{\partial u_+(f)}{\partial \nu}\right|_{\partial\Omega} \qquad (f \in H^{1/2}(\partial\Omega)),$$

where $u_+ = u_+(f) \in H^1(B(R)\backslash\overline{\Omega})$ is the solution to

$$\begin{cases} (\Delta + k_0^2)u_+ = 0 & \text{in } B(R)\backslash\overline{\Omega}, \\ \text{b.c.,} \\ u_+|_\Gamma = 0, \quad u_+|_{\partial\Omega} = f. \end{cases}$$

Also, define the trace K of the single layer potential operator by

$$Kf := u(f)|_{\partial\Omega} \qquad (f \in H^{-1/2}(\partial\Omega)),$$

where $u = u(f) \in H^1(B(R))$ is the solution to

$$\begin{cases} (\Delta + k_0^2 n^2(\mathbf{x}))u = -M_f & \text{in } B(R), \\ \text{b.c.,} \\ u|_\Gamma = 0, \end{cases}$$

with $M_f \in H^{-1}(B(R))$ defined by

$$M_f(\phi) = \langle f, \phi|_{\partial\Omega}\rangle \qquad (\phi \in \dot{H}^1(B(R))).$$

Then, likewise the previous Lemma 9.5, we have the following lemma.

Lemma 9.7. (i) $\Lambda^- - \Lambda^+$ *is injective.*
(ii) $(\Lambda^- - \Lambda^+)K = I.$

So, we know Λ^- if we know K.
Next we will see that K can be known from Λ by the following two lemmas:

Lemma 9.8. *Let $u = u(f)$ be the solution used to define Kf. For $f_0 \in H^{-1}(B(R))$
with supp $f_0 \subset B(R)\backslash\overline{\Omega}$, let $u_0 = u_0(f_0) \in H^1(B(R))$ be the solution to*

$$\begin{cases} (\Delta + k_0^2)u_0 = -f_0 & in\ B(R), \\ b.c., \\ u_0|_\Gamma = 0. \end{cases}$$

Also, let $w \in H^1(B(R))$ be the solution to

$$\begin{cases} (\Delta + k_0^2 n^2)w = -k_0^2(n^2 - 1)u_0 & in\ B(R), \\ b.c., \\ w|_\Gamma = 0. \end{cases}$$

Then, we have

$$f_0(u) = \langle f, u_0|_{\partial\Omega}\rangle + \langle f, w|_{\partial\Omega}\rangle.$$

So, we know K if we know w.
The next lemma shows that we know w from Λ.

Lemma 9.9. *Let $g_0 \in H^{-1}(B(R))$ satisfy supp $g_0 \subset B(R)\backslash\overline{\Omega}$. Then, we have*

$$-g_0(w) = \lim_{j\to\infty} \langle(\Lambda - \Lambda_0)u_j|_\Gamma, v_j|_\Gamma\rangle,$$

where $u_j, v_j \in H^1(B(R))$ satisfy

$$\begin{cases} (\Delta + k_0^2)u_j = 0 & in\ B(R), \\ b.c., \\ u_j \to u_0 & in\ H^1(\Omega)\ (j \to \infty), \end{cases}$$

$$\begin{cases} (\Delta + k_0^2)v_j = 0 & in\ B(R), \\ b.c., \\ v_j \to v_0 & in\ H^1(\Omega)\ (j \to \infty), \end{cases}$$

u_0 is given in Lemma 9.8 and $v_0 \in H^1(B(R))$ is the solution to

$$\begin{cases} (\Delta + k_0^2)v_0 = -g_0 & in\ B(R), \\ b.c., \\ v_0|_\Gamma = 0. \end{cases}$$

Remarks. (i) The existence of u_j, v_j in Lemma 9.9 follows from the following
Runge-type theorem.
Put $L_0 = \Delta + k_0^2$. Assume that $B(R)\backslash\overline{\Omega}$ is connected. Define the sets X, Y by

$$X = \{u|_\Omega : u \in H^1(\Omega(u)), L_0 u = 0 \text{ in } \Omega(u)\},$$
$$Y = \{v|_\Omega : u \in H^1(\Omega(v)), L_0 v = 0 \text{ in } \Omega(v), \text{ b.c.}\},$$

respectively. Here, for any u, $\Omega(u)$ is an open set depending on u such that

$$\Omega \subset \overline{\Omega} \subset \Omega(u) \subset \overline{\Omega(u)} \subset B(R).$$

Then, Y is dense in X with respect to the $H^1(\Omega)$ norm.

(ii) The Nachman reconstruction procedure is as follows. Fix any $\mathbf{k} \in \mathbb{R}^3$ and let $\zeta \in \mathbb{C}^3$ satisfy $\zeta \cdot \zeta = (\zeta + \mathbf{k}) \cdot (\zeta + \mathbf{k}) = k_0^2$. Define $t(\mathbf{k}, \zeta)$ by

$$t(\mathbf{k}, \zeta) := \langle (\Lambda^- - \Lambda^{-,0}) e^{-\sqrt{-1}\mathbf{x} \cdot (\zeta + \mathbf{k})}|_{\partial\Omega}, \left(\frac{1}{2} I + S_\zeta \Lambda^- - W_\zeta \right)^{-1} e^{\sqrt{-1}\mathbf{x} \cdot \zeta}|_{\partial\Omega} \rangle,$$

where S_ζ and W_ζ are the traces of the single layer potential and double layer potential of $G_\zeta := e^{\sqrt{-1}\mathbf{x} \cdot \zeta} (\Delta + 2\sqrt{-1}\zeta \cdot \nabla)^{-1}$ to $\partial\Omega$, and $\Lambda^{-,0}$ is Λ^- with $n = 1$. Then,

$$\lim_{|\zeta| \to \infty} t(\mathbf{k}, \zeta) = -\int_\Omega e^{-\sqrt{-1}\mathbf{x} \cdot \mathbf{k}} k_0^2 (n^2 - 1)\, dx.$$

9.4   References

[C]   K. Chelminski. The principle of limiting absorption in elasticity. *Bull. Polish Acad. Sci. Math.*, **41**:219–230, 1993.

[CD1] E. Croc and Y. Dermenjian. Spectral analysis of a multistratified acoustic strip. I. Limiting absorption principle for a simple stratification. *SIAM J. Math. Anal.*, **26**:880–924, 1995.

[CD2] E. Croc and Y. Dermenjian. Spectral analysis of a multistratified acoustic strip. II. Asymptotic behavior of solutions for a simple stratification. *SIAM J. Math. Anal.*, **27**:1631–1652, 1996.

[CD3] E. Croc and Y. Dermenjian. A perturbative method for the spectral analysis of an acoustic multistratified strip. *Math. Methods Appl. Sci.*, **21**:1681–1704, 1998.

[GX] R.P. Gilbert and Y. Xu. The propagation problem and far-field patterns in a stratified finite depth ocean. *Math. Methods Appl. Sci.*, **12**:199–208, 1990.

[I]   M. Ikehata. Reconstruction of inclusion from boundary measurements (preprint).

[IMN] M. Ikehata, G. Makarakis and G. Nakamura. Inverse boundary value problem for ocean acoustics. *Math. Methods Appl. Sci.*, to appear.

[M]   S. Mizohata. *The Theory of Partial Differential Equations*. Cambridge University Press, London, 1973.

[N]   A. Nachman. Reconstruction from boundary measurements. *Ann. of Math.*, **128**:531–577, 1988.

[X]   Y. Xu. The propagating solution and far-field patterns for ocean acoustic harmonic waves in a finite depth ocean. *Appl. Anal.*, **35**:129–151, 1990.

10

Multidimensional Inverse Problem for the Acoustic Equation in the Ray Statement

Valery G. Yakhno

ABSTRACT This chapter deals with a multidimensional inverse dynamic problem for the acoustic equation. This problem is reduced to inverse kinematic, integral geometry problems, a Dirichlet problem for a quasilinear elliptic equation, and for some special case to standard tomography problems. The structure and some properties of the fundamental solution of the Cauchy problem for the acoustic equation are used for this reduction. The structure and properties are described in detail.

10.1 Introduction

Let us consider the propagation of the front of an acoustic wave in the homo-geneous acoustic medium contained in the three—dimensional space. The front shape and the velocity of its points depend on the velocity of the acoustic wave inside this medium. In the case when the velocity depends on all three coordi-nates the wavefront can be represented by an intricate surface. Let D be a domain of three-dimensional space $\mathbf{R}^3$ limited by a boundary S, wherein transfer of sig-nals with the finite positive velocity $v(x)$, $x = (x_1, x_2, x_3) \in \mathbf{R}^3$ takes place. Let $\tau(x, x^0)$ be the time required for the signal to get from the point x^0 to point x. The function $\tau(x, x^0)$ satisfies the eikonal equation

$$\left|\nabla_x \tau(x, x^0)\right|^2 = n(x), \qquad n(x) = \frac{1}{v^2(x)}, \qquad (10.1)$$

and the condition

$$\tau(x, x^0) = O\left(|x - x^0|\right), \qquad |x - x^0| \to 0. \qquad (10.2)$$

Here ∇_x is the gradient calculated with respect to the variale x. Equation $\tau(x, x^0) = t$ defines the wavefront from the point source at x^0 at time t. Therefore, the problem of finding $v(x)$ by the law of the front motion along the surface S can be set as follows:

Problem 10.1 (Inverse Kinematic Problem). Find $v(x)$ for $x \in D$, if we know the function $\tau(x, x^0)$ for $x \in S$ and $x^0 \in S$, which is a solution of (10.1) under condition (10.2).

The solution of the inverse kinematic problem for the spherically symmetrical model of the Earth was studied by Herglotz [HR] and Wiechert [WK] at the beginning of the twentieth century. It is worth saying that this inverse kinematic problem was the first one considered for partial differential equations. Results of investigations of the inverse kinematic problem can be found in the works [GM], [AN], [MR], [BE], [BG], and others. We note that the eikonal equation (10.1) describes the travel-time of the acoustic wave propagation only, but dynamics of the acoustic field may be described by the following acoustic equation:

$$\frac{\partial^2 u}{\partial t^2} = v^2(x)\Delta u - v^2(x)\nabla_x \ln m(x)\nabla_x u + \sigma(x)\frac{\partial u}{\partial t} + f(x, t). \tag{10.3}$$

Here $x = (x_1, x_2, x_3) \in \mathbf{R}^3$, $t \in \mathbf{R}$; Δ_x is the Laplace operator with respect to the variable x. The physical meaning of $u(x, t)$ is a pressure of the acoustic medium at the point x at the moment t; $f(x, t)$ is the density of the externally acting forces (sources); $v(x)$ is the velocity function, $v(x) > 0$; $m(x)$ is the density function, $m(x) > 0$; and $\sigma(x)$ is the attenuation function. The coefficients of the acoustic equation $v(x)$, $m(x)$, $\sigma(x)$ are the characteristics of the acoustic medium.

There are a lot of applications where the following inverse problem may be interesting:

Problem 10.2 (Inverse Dynamic Problem). Let $f = \delta(x - x^0, t)$ be the Dirac delta function; let D be a given domain in $\mathbf{R}^3$ with the smooth boundary S; and let the functions $v(x), m(x), \sigma(x)$ be unknown for $x \in D$ and have the constant values for $x \in \mathbf{R}\backslash D$. One has to find these functions (from differential equation (10.3)) if the following information, about the solution (10.3) subject to the condition

$$u\big|_{t<0} = 0, \tag{10.4}$$

is given

$$u(x, t, x^0)\big|_{x \in S, x^0 \in S,} = G(x, t, x^0), \tag{10.5}$$

where $x^0 = (x_1^0, x_2^0, x_3^0)$ is a parameter of the problem; $\delta(x - x^0, t)$ means a pulse point source concentrating at the point $x = x^0$ at the moment $t = 0$; $G(x, t, x^0)$ is the acoustic wave field which is known for arbitrary $x \in S$ and $x^0 \in S$ and for every t from $[0, T]$; and T is a sufficiently large positive number.

We note that this inverse problem is the natural extension of the inverse kinematic problem because we wish to find here all the unknown characteristics of a nonhomogeneous medium instead of finding the velocity function in the inverse kinematic problem. The other inverse problems for acoustic equations the reader can find in [LRS], [ROM1], [ROMK], [YA1].

This chapter discovers a method of all function-coefficients recovering using dynamic information of an acoustic wave field. This dynamic information is the supplementary data about wave fields around arrivals of the front arising from the pulse point source at an arbitrary point $x^0 \in S$. Reduction of this inverse problem to the inverse kinematic problem and two integral geometry problems, and the Dirichlet problem for the quasilinear elliptic equation is given in this chapter.

The chapter has the following structure. The first two sections describe concepts, operations, and properties from partial differential equations and generalized function theory which we actively use in the following sections. The important nonstandard properties and propositions are provided by proofs, the others by brief description only. We use in our text some concepts, notations, and elements from the books [ROM1] and [VL]. Sections 10.4 and 10.5 present the existence theorem of the fundamental solution of the Cauchy problem for the acoustic equation. The properties of this fundamental solution are clarified here also. Section 10.6 contains the description of inverse problem solving. Some remarks about particular cases, generalizations, and open questions for investigations may be found in the last section.

10.2   Acoustic, Eiconal, Euler's Equations, and Rays

The aim of this section is to bring notions of acoustic, eiconal, Euler's equations, rays, and the Riemmann coordinates, and to describe their relationships and some properties which we use in our investigation. Notations and a description of this section is similar in spirit to Romanov's book [ROM1].

Let us consider for $x = (x_1, x_2, x_3) \in \mathbf{R}^3, t \in \mathbf{R}$, the acoustic equation (10.3) for $f = \delta(x - x^0, t)$, where $x^0 = (x_1^0, x_2^0, x_3^0) \in \mathbf{R}^3$ is a parameter. The functions $v(x)$, $m(x)$, $\sigma(x)$ appearing in the acoustic equation (10.3) as coefficients are given. We describe now the main assumptions over these function–coefficients of the acoustic equation. There are two types of assumptions. The first of them is the following:

Assumption 10.3 (A1).   All functions $v(x)$, $m(x)$, $\sigma(x)$ are infinitely differentiable functions of their arguments and

$$m(x) > 0, \qquad 0 < v_1 \le v(x) \le v_2,$$

where v_1, v_2 are given positive constants.

To formulate the second-type assumptions we need to introduce some notions.

Notion 10.4 (The Characteristic Surface of the Acoustic Equation).   The equation of a characteristic surface of the acoustic equation is given in the form $t = \tau(x)$, where function $\tau(x)$ satisfies the eikonal equation (10.1).

Among all possible solutions of (10.1) there are such solutions that form characteristic surfaces $t = \tau(x, x^0)$ having a conic point at an arbitrary fixed point x^0. Such surfaces are characteristic conoids. To find the function $\tau(x, x^0)$, equation (10.1) must be integrated under the condition (10.2). The method in [EL] of constructing characteristic conoids $t = \tau(x, x^0)$ consists in constructing separate lines, called bicharacteristics, that lie on the conoid and jointly form it. Formally, the procedure of obtaining the solution is as follows: consider the vector

$$p = \nabla_x \tau(x, x^0), \qquad p = (p_1, p_2, p_3), \tag{10.6}$$

and from the eikonal equation (10.1), which can be represented in the form $|p|^2 = n^2(x)$, by differentiating with respect to x_k, we get the equation

$$p\frac{\partial p}{\partial x_k} = n\frac{\partial n}{\partial x_k}, \qquad k = 1, 2, 3. \tag{10.7}$$

Then, using the equality

$$\frac{\partial p_i}{\partial x_k} = \frac{\partial p_k}{\partial x_i},$$

resulting from (10.6), we can write (10.7) as follows:

$$p\nabla_x p_k = n\frac{\partial n}{\partial x_k}, \qquad k = 1, 2, 3.$$

Thus each component of the vector p satisfies the first-order quasilinear equation. Along the curves satisfying the equality

$$\frac{dx}{dt} = \frac{p}{n^2(x)}$$

equation (10.7), after being divided into $n^2(x)$, can be written in the following way:

$$\frac{dp_k}{dt} = \frac{\partial \ln n(x)}{\partial x_k}, \qquad k = 1, 2, 3,$$

and $\tau(x, x^0)$ satisfies the equation

$$\frac{d\tau}{dt} = \nabla_x \tau \frac{dx}{dt} = \frac{|p|^2}{n^2(x)} = 1.$$

By choosing the parameter t in such way that $\tau = 0$ at $t = 0$, we get $\tau = t$, i.e., the parameter t is the time of the signal transfer from the point x^0 to the point x. The system of equalities

$$\frac{dx}{dt} = \frac{p}{n^2(x)}, \qquad \frac{dp}{dt} = \nabla_x \ln n(x) \tag{10.8}$$

is the ordinary differential equation with respect to the pair of unknown vector-functions x, p.

Notion 10.5 (Euler's System).   System (10.8) is called Euler's system.

Assume that

$$x|_{t=0} = x^0, \qquad p|_{t=0} = p^0 \equiv n(x^0)v^0, \tag{10.9}$$

where v^0 is an arbitrary unit vector. By way of solving (10.8) and (10.9) we find x, p as functions of t and parameters x^0, p^0:

$$x = x(t, x^0, p^0), \qquad p = p(t, x^0, p^0). \tag{10.10}$$

The first of equalities (10.10) determines a two-parametric set of bicharacteristics, forming a characteristic conoid in the space of variables x, t at a fixed x^0.

Notion 10.6 (Ray).   Projection of the bicharacteristic onto the space of x is called a ray.

The equality $x = x(t, x^0, p^0)$ can be considered as a parametric formulation of this ray. The first of the relations (10.8) demonstrates that a tangent to the ray coincides with the direction of the vector p and, due to (10.6), the rays are orthogonal to the surfaces $\tau(x, x^0) = t$.

As is known from the calculus of variation [EL] that rays are extremals of the functional

$$\tau(L) = \int_{L(x^0, x)} n(x)\, ds,$$

where $L(x^0, x)$ is an arbitrary smooth curve connecting a pair of points x^0, x; ds is an element of its length. The product $n(x)ds$ determines the elementary time required for the signal to pass the length ds with the velocity $v(x)$. The integral over the curve $L(x^0, x)$ has, therefore, a physical sense of the time spent by the signal to pass along the curve $L(x^0, x)$ from point x^0 to the point x.

Notion 10.7 (Extremals).   Those curves on which this integral reaches extremum are called the extremals of the functional $\tau(L)$.

As is demonstrated in the calculus of variation, the functional extremals satisfy Euler's system (10.8). Thus, the rays introduced as the projections of the bicharacteristics of the eikonal equation onto the space of x are, in fact, the extremals of the functional $\tau(L)$.

Our second assumption about coefficients of the acoustic equation is the following:

Assumption 10.8 (A2).   The family of rays $\{T(x^0, x)\}|_{x^0 \in \mathbf{R}^3, x \in \mathbf{R}^3}$ is regular.

This means that for each pair of points x^0, x from $\mathbf{R}^3$ there exists a ray $T(x^0, x)$ which connects these points and this ray is unique. We note that the assumption A2 is equivalent to the following requirement:

Assumption 10.9 (A2a).   For each pair of points x^0, x from $\mathbf{R}^3$ there exists an extremal of the functional $\tau(L)$ connecting these points x^0, x and this extremal is unique.

Let now $x = x(t, x^0, p^0)$, $p = p(t, x^0, p^0)$ be a solution of (10.8), (10.9) and $\zeta = v^0 v(x^0)t$, $n(x) = 1/v(x)$. Then we can write the following representation

$$x = f(\zeta, x^0), \qquad p = \frac{n^2(f(\zeta, x^0))}{t}\zeta \frac{\partial f(\zeta, x^0)}{\partial \zeta},$$

where the vector-function $f(\zeta, x^0)$, $f = (f_1, f_2, f_3)$ has continuous derivatives over its arguments; $\partial f / \partial \zeta$ is the Jacobian matrix

$$\frac{\partial f}{\partial \zeta} = \begin{pmatrix} \frac{\partial}{\partial \zeta_1} f_1 & \frac{\partial}{\partial \zeta_1} f_2 & \frac{\partial}{\partial \zeta_1} f_3 \\ \frac{\partial}{\partial \zeta_2} f_1 & \frac{\partial}{\partial \zeta_2} f_2 & \frac{\partial}{\partial \zeta_2} f_3 \\ \frac{\partial}{\partial \zeta_3} f_1 & \frac{\partial}{\partial \zeta_3} f_2 & \frac{\partial}{\partial \zeta_3} f_3 \end{pmatrix}, \qquad \zeta = (\zeta_1, \zeta_2, \zeta_3).$$

Differential properties of the function $f(\zeta, x^0)$ result from the corresponding properties of the function $x = x(t, x^0, p^0)$. The point $\zeta = 0$ is conformed to the value of the parameter $t = 0$. Let us write the first terms of the function $f(\zeta, x^0)$ extension into the Taylor series in the vicinity of this point. Using the Cauchy data and equality (10.8) one finds

$$x = x^0 + \frac{dx}{dt}\Big|_{t=0} t + O(t^2) = x^0 + \frac{p^0 t}{n^2(x^0)} + O(t^2).$$

Hence

$$f(\zeta, x^0) = x^0 + \zeta + O(|\zeta^2|).$$

The assumption A2 (or A2a) is equivalent to the following:

Assumption 10.10 (A2b).   For each pair of points ζ, x^0 from $\mathbf{R}^3$ the inequality

$$\left| \frac{\partial f(\zeta, x^0)}{\partial \zeta} \right| \equiv \det\left(\frac{\partial f(\zeta, x^0)}{\partial \zeta} \right) \neq 0$$

holds, and the function $f(\zeta, x^0)$ has the inverse one, $\zeta = g(x, x^0)$, for which $g(x^0, x^0) = 0$, and

$$\left| \frac{\partial g(x, x^0)}{\partial x} \right| = \left[\left| \frac{\partial f(\zeta, x^0)}{\partial \zeta} \right| \Big|_{\zeta = g(x, x^0)} \right]^{-1},$$

$$\left| \frac{\partial g(x, x^0)}{\partial x} \right| \Big|_{x = x^0} = 1.$$

Under the suppositions A1 and A2 (or A1, A2a, or A1, A2b) and at the point x^0 fixed, the point x can be given with the Riemann coordinates $\zeta = (\zeta_1, \zeta_2, \zeta_3)$, i.e., $x = f(\zeta, x^0)$.

Notion 10.11 (Riemann's Coordinates).   The Riemann coordinates have the following sense: $\zeta = \tau(x, x^0)\alpha$, where $\alpha = (\alpha_1, \alpha_2, \alpha_3)$ is a vector directed at the x^0 along the tangent to the ray $T(x^0, x)$ toward the point x and such that $|\alpha|^2 = v^2(x^0)$.

Jacobian of transition from the Riemann to the Cartesian coordinates is

$$\left| \frac{\partial g(x, x^0)}{\partial x} \right| \equiv \det\left(\frac{\partial g(x, x^0)}{\partial x} \right) \neq 0,$$

and at $x = x^0$ is equal to 1.

Let x^0 be a fixed point from $\mathbf{R}^3$; and let $\alpha = (\alpha_1, \alpha_2, \alpha_3)$ be an arbitrary unit vector. Let us consider a ray which goes through x^0 and has the tangent vector at the direction of α. Let x be an arbitrary point of this ray, then this point can be given with the Riemann coordinates $x = f(\zeta, x^0)$, where $\zeta = \alpha\tau$, $(\zeta = (\zeta_1, \zeta_2, \zeta_3))$. Here τ is a parameter such that $\tau = \tau(x, x^0)$. The function $x(\tau) = f(\alpha\tau, x^0)$ is an integral of system

$$\frac{dx}{d\tau} = v^2(x)\nabla_x\tau(x, x^0). \tag{10.11}$$

There is the following property which is important for us. It can be formulated as the following lemma:

Lemma 10.12. *Let $\tau(x, x^0)$ be a solution of (10.1), (10.2); $\Gamma = t^2 - \tau^2(x, x^0)$. Then under the requirements A1, A2 along a ray which goes through x^0 and has tangent vector at the direction of α the following equality holds:*

$$\mathrm{div}\,(v^2(x)\nabla_x\Gamma)\big|_{x=x(\tau)} = -6 + 2\tau\frac{\partial}{\partial\tau}\left[\ln\left|\frac{\partial g(x, x^0)}{\partial x}\right|\bigg|_{x=x(\tau)}\right].$$

PROOF.   Denoting $v^2(x)\nabla_x\Gamma = F(x, x^0)$ and using the above-mentioned notations and (10.11), we have along the curve $x = f(\zeta, x^0)$ the following equality:

$$\frac{\partial f(\zeta, x^0)}{\partial\zeta}\zeta = -\tfrac{1}{2}v^2(f(\zeta, x^0))\nabla_x\Gamma.$$

Hence

$$-2\frac{\partial f(\zeta, x^0)}{\partial\zeta}\zeta = F(f(\zeta, x^0), x^0).$$

Therefore the following relations are valid:

$$\tau\frac{\partial}{\partial\tau}\left(\frac{\partial f_i(\zeta, x^0)}{\partial\zeta_k}\right) = \frac{\partial}{\partial\zeta}\left(\frac{\partial f_i(\zeta, x^0)}{\partial\zeta_k}\right)\zeta$$

$$= \frac{\partial}{\partial\zeta}\left(\frac{\partial f_i(\zeta, x^0)}{\partial\zeta_k}\zeta\right) - \frac{\partial f_i(\zeta, x^0)}{\partial\zeta_k}$$

$$= -\frac{1}{2}\sum_{j=1}^{3}\frac{\partial}{\partial x_j}F_i(f(\zeta, x^0), x^0)\frac{\partial f_j(\zeta, x^0)}{\partial\zeta_k} - \frac{\partial f_i(\zeta, x^0)}{\partial\zeta_k}. \tag{10.12}$$

Calculating

$$\tau\frac{\partial}{\partial\tau}\left|\frac{\partial f}{\partial\zeta}\right|$$

by way of differentiating the determinant with respect to the columns and using
(10.12) we have

$$
\tau \frac{\partial}{\partial \tau} \left| \frac{\partial f}{\partial \zeta} \right| = - \left| \frac{\partial f(\zeta, x^0)}{\partial \zeta} \right| \left(\frac{1}{2} \sum_{j=1}^{3} \frac{\partial F_i(x, x^0)}{\partial x_i} + 3 \right) \Bigg|_{x=f(\zeta, x^0)}.
$$

Using the assumption A2b we obtain, as the result,

$$
\mathrm{div}_x \, F(x, x^0)|_{x=f(\zeta, x^0)} = -6 + 2\tau \frac{\partial}{\partial \tau} \left[\ln \left| \frac{\partial g(x, x^0)}{\partial x} \right| \Bigg|_{x=f(\zeta, x^0)} \right]
$$

and the lemma is proved.

10.3   Basic Facts from Generalized Functions

In this section we look very briefly at the definition of generalized functions and
indicate some their properties which can be found in book [VL] and prove other
ones that are not standard.

Definition 10.13 (The Space of Test Functions **D**).   Let us relate to the test func-
tions $\mathbf{D} = \mathbf{D}(\mathbf{R}^n)$ all the infinitely differentiable functions in $\mathbf{R}^n$ with compact
support. We shall define convergence in **D** as follows. The sequence of the functions
$\phi_1, \phi_2, \ldots$, from **D** converges to the function ϕ (belonging to **D**) if:

(i) there exists a number $M > 0$ such that $\mathrm{supp}\,\phi_k \subset \{x \in \mathbf{R}^n \mid |x| \leq M\}$;

(ii) for each $\alpha = (\alpha_1, \alpha_2, \ldots, \alpha_n)$ with integer nonnegative components α_j the
following relation holds:

$$
\sup_{x \in \mathbf{R}^n} |D^\alpha \phi_k(x) - D^\alpha \phi(x)| \to 0 \quad \text{as} \quad k \to \infty.
$$

Here $\mathrm{supp}\,\phi_k$ is a closure of the set $\{x \in \mathbf{R}^n \mid \phi_k(x) \neq 0\}$:

$$
D^\alpha = \frac{\partial^{|\alpha|}}{\partial x_1^{\alpha_1} \partial x_2^{\alpha_2} \ldots \partial x_n^{\alpha_n}}, \qquad |\alpha| = \alpha_1 + \alpha_2 + \cdots + \alpha_n.
$$

Definition 10.14 (The Space of Generalized Functions **D**′).   Each linear continuous functional over the space of test functions **D** is known as a generalized
function in the Sobolev–Schwartz sense. We shall denote by (f, ϕ) the effect of
the functional (generalized function) f over the test function ϕ. We shall also
formally denote the generalized function f by $f(x)$, with an understanding that x
is the argument of the test functions on the functional acts.

Let us interpret the definition of the generalized function:

(1) The generalized function is a functional over **D**. That is, a (complex) number
(f, ϕ) is associated with each $\phi \in \mathbf{D}$.

(2) The generalized function f is a linear functional over $\mathbf{D}$; that is, if $\phi \in \mathbf{D}$, $\psi \in \mathbf{D}$, and λ, μ are complex numbers, then

$$(f, \lambda\phi + \mu\psi) = \lambda(f, \phi) + \mu(f, \psi).$$

(3) The generalized function f is a continuous functional over $\mathbf{D}$. That is, if $\phi_k \to \phi$ as $k \to \infty$ in $\mathbf{D}$, then

$$(f, \phi_k) \to (f, \phi) \qquad \text{as} \quad k \to \infty.$$

We shall denote the set of all generalized functions by $\mathbf{D}' = \mathbf{D}'(\mathbf{R}^n)$. The set $\mathbf{D}'$ is linear. We shall define convergence in $\mathbf{D}'$ in the following way. The sequence of generalized functions $f_1, f_2, \ldots, f_k, \ldots$, belonging to $\mathbf{D}'$ converges to the generalized function $f \in \mathbf{D}'$, if $(f_k, \phi) \to (f, \phi)$ as $k \to \infty$. The linear set $\mathbf{D}'$ with such a convergence is known as the space of generalized functions $\mathbf{D}'$. We note that this space $\mathbf{D}'$ is complete.

The generalized function f becomes zero in the region G if $(f, \phi) = 0$ for all $\phi \in \mathbf{D}(\mathbf{R}^n)$ such that $\operatorname{supp} \phi \subset G$. We shall write this fact thus: $f = 0$ for $x \in G$. In correspondence with this definition, the generalized functions f and g are said to be equal in the region G if $f - g = 0$ for all $x \in G$; in this case we write $f = g$ for $x \in G$. Specifically, the generalized functions f and g are said to be equal, $f = g$, if for all $\phi \in \mathbf{D}$, $(f, \phi) = (g, \phi)$. We shall say that the generalized function f belongs to the class $C^p(G)$ if in the region G it coincides with the function $f_0(x)$ of the class $C^p(G)$; that is, for any $\phi \in \mathbf{D}(G)$:

$$(f, \phi) = \int_{\mathbf{R}^n} f_0(x)\phi(x)\, dx.$$

Definition 10.15 (Regular Generalized Functions). Generalized functions which are definable in terms of functions locally integrable in $\mathbf{R}^n$ according to formula

$$(f, \phi) = \int_{\mathbf{R}^n} f(x)\phi(x)\, dx, \qquad f(x) \in L^{\mathrm{loc}}(\mathbf{R}^n), \qquad \phi \in \mathbf{D}(\mathbf{R}^n) \qquad (10.13)$$

are said to be regular generalized functions.

Definition 10.16 (Singular Generalized Functions). The remaining generalized functions are said to be singular generalized functions.

There is the following very important lemma:

Lemma 10.17 (Du Bois Reymond [VL]). *In order that the function $f(x)$, locally integrable in G, should become zero in the region G in the sense of generalized functions, it is necessary and sufficient that $f(x) = 0$ almost everywhere in G.*

Each function locally integrable in $\mathbf{R}^n$ defines according to formula (10.13) a regular generalized function. It follows from Du Bois Reymond's lemma that each regular generalized function is defined by a unique (with an accuracy as far as values on a set of measure zero) function locally integrable in $\mathbf{R}^n$. Consequently, there is a mutual one-to-one correspondence between functions locally integrable in $\mathbf{R}^n$ and regular generalized functions. Therefore we shall identify a locally integrable function $f(x)$ and a generalized function generated by it according to formula (10.13). In this sense the "usual," that is, locally integrable functions in $\mathbf{R}^n$, are (regular) generalized functions.

By the definition we have just given, it is impossible to identify a singular generalized function with any locally integrable function. The simplest example of a singular generalized function is the Dirac delta-function

$$(\delta, \phi) = \phi(0), \qquad \phi \in \mathbf{D},$$

(see, for instance, [VL]).

Let us now consider some important operations over generalized functions.

Operation 10.18 (Change of Variables in Generalized Functions). Let $f(y)$ be a generalized function from $\mathbf{D}'(\mathbf{R}^n)$, let $y = \omega(x)$ be an infinitely differentiable one-to-one transformation of $\mathbf{R}^n$ onto itself with nonzero determinant of the Jacobian $\partial \omega / \partial x$, i.e.,

$$\det \left(\frac{\partial \omega(x)}{\partial x} \right) = \left| \frac{\partial \omega(x)}{\partial x} \right| \neq 0;$$

and let $x = \omega^{-1}(y)$ be the inverse transformation to $y = \omega(x)$. Then, for any $\phi \in \mathbf{D}$ the relation

$$\left(f(\omega(x)), \phi(x) \right) = \left(f(y), \phi(\omega^{-1}(y)) \left| \frac{\partial \omega^{-1}(y)}{\partial y} \right| \right). \tag{10.14}$$

defines the generalized function $f(\omega(x))$ for any $f(y) \in \mathbf{D}'(\mathbf{R}^n)$.

For instance, using this definition we can show the following properties of the Dirac delta function

$$(\delta(x - x^0), \phi(x)) = \phi(x^0), \qquad \phi \in \mathbf{D}(\mathbf{R}^n), \tag{10.15}$$

$$\delta(-x) = \delta(x). \tag{10.16}$$

If $\omega^{-1}(0) = x^0$, then

$$\delta(\omega(x)) = \frac{\delta(x - x^0)}{|\partial \omega(x)/\partial x| \big|_{x=x^0}}. \tag{10.17}$$

Operation 10.19 (Multiplication of Generalized Functions). Let $f(x) \in \mathbf{D}'(\mathbf{R}^n)$, $a(x) \in C^\infty(\mathbf{R}^n)$. The equation

$$(af, \phi) = (f, a\phi), \qquad \phi \in \mathbf{D},$$

defines the product af for any $f(x) \in \mathbf{D}'(\mathbf{R}^n)$.

Using this definition we can obtain the following properties of the Dirac delta function

$$a(x)\delta(x) = a(0)\delta(x), \qquad a(x)\delta(x - x^0) = a(x^0)\delta(x - x^0). \tag{10.18}$$

Operation 10.20 (Differentiation of Generalized Functions). Let $f(x) \in$ $\mathbf{D}'(\mathbf{R}^n)$, and let $\alpha = (\alpha_1, \alpha_2, \ldots, \alpha_n)$ be a vector with integer nonnegative components α_j; $|\alpha| = \alpha_1 + \alpha_2 + \cdots + \alpha_n$:

$$D^\alpha f(x) = \frac{\partial^{|\alpha|} f(x_1, x_2, \ldots, x_n)}{\partial x_1^{\alpha_1} \partial x_2^{\alpha_2} \ldots \partial x_n^{\alpha_n}}, \qquad D^0 f(x) = f(x).$$

We take the equation

$$(D^\alpha f(x), \phi) = (-1)^{|\alpha|}(f, D^\alpha \phi(x)), \qquad \phi \in \mathbf{D},$$

as the definition of the (generalized) derivative $D^\alpha f$ of the generalized function $f(x) \in \mathbf{D}'(\mathbf{R}^n)$.

From the definition of the generalized derivative the following properties hold (see, [VL]):

- If $f(x) \in \mathbf{D}'(\mathbf{R}^n)$ then $D^\alpha f \in \mathbf{D}'(\mathbf{R}^n)$ for arbitrary α.

- If $f(x) \in C^p(G)$, $\{D^\alpha f\}$ is the classical derivative (where it exists), then $D^\alpha f = \{D^\alpha f(x)\}$, $x \in G$, $|\alpha| \leq p$.

- Any generalized function is infinitely differentiable.

- The result of differentiation does not depend on the order of differentiation.

- If $f(x) \in \mathbf{D}'(\mathbf{R}^n)$ and $a \in C^\infty(\mathbf{R}^n)$, then Leibnitz's formula for differentiation of the product af is valid.

- If the generalized function $f = 0$ for $x \in G$, then also $D^\alpha f = 0$ for $x \in G$, so supp $D^\alpha f \subset$ supp f.

- If $t \in \mathbf{R}$ and the function $f(t)$ has isolated discontinuities of the first kind at the points $\{t_k\}$, then

$$\frac{df}{dt} = \left\{ \frac{df}{dt} \right\} + \sum_k [f] \Big|_{t_k} \delta(t - t_k), \qquad (10.19)$$

where we write $[f]|_{t_k} = f(t_k + 0) - f(t_k - 0)$.

- If

$$\theta(t) = \begin{cases} 1, & t \geq 0 \\ 0, & t < 0 \end{cases} \qquad \text{the Heaviside function,}$$

then

$$\frac{d\theta(t)}{dt} = \delta(t), \qquad (10.20)$$

$$t^n \delta^n(t) = (-1)^n n! \, \delta(t). \qquad (10.21)$$

The following lemma holds (see [VL]):

Lemma 10.21. *For any $g(y) \in \mathbf{D}'(\mathbf{R}^m)$, and any $\phi(x, y) \in \mathbf{D}(\mathbf{R}^{n+m})$ the function*

$$\psi(x) = (g(y), \phi(x, y)) \in \mathbf{D}(\mathbf{R}^n).$$

Moreover, for all α:

$$D^\alpha \psi(x) = (g(y), D^\alpha \phi(x, y)).$$

Further, if $\phi_k \to \phi$ as $k \to \infty$ in $\mathbf{D}(\mathbf{R}^{n+m})$, then

$$\psi_k(x) = (g(y), \phi_k(x, y)) \to \psi(x), \qquad k \to \infty \qquad in \ \mathbf{D}(\mathbf{R}^n).$$

Operation 10.22 (The Direct Product of Generalized Functions). Let $f(x) \in \mathbf{D}'(\mathbf{R}^n)$, $g(y) \in \mathbf{D}'(\mathbf{R}^m)$, then the equality

$$(f(x) \cdot g(y), \phi) = (f(x), (g(y), \phi(x, y))), \qquad \phi(x, y) \in \mathbf{D}(\mathbf{R}^{n+m}),$$

defines the direct product $f(x) \cdot g(y)$ of the generalized functions $f(x) \in \mathbf{D}'(\mathbf{R}^n)$, $g(y) \in \mathbf{D}'(\mathbf{R}^m)$.

The following properties of the direct product hold (see [VL]):

- *Commutativity*: $f(x) \cdot g(y) = g(y) \cdot f(x)$.
- *Continuity*: If $f_k \to f$ as $k \to \infty$ in $\mathbf{D}'(\mathbf{R}^n)$, then $f_k(x) \cdot g(y) \to f(x) \cdot g(y)$ as $k \to \infty$ in $\mathbf{D}'(\mathbf{R}^{n+m})$.
- *Associativity*: If $f(x) \in \mathbf{D}'(\mathbf{R}^n)$, $g(y) \in \mathbf{D}'(\mathbf{R}^m)$, and $h(z) \in \mathbf{D}'(\mathbf{R}^k)$, then $f(x) \cdot (g(y) \cdot h(z)) = (f(x) \cdot g(y)) \cdot h(z)$.
- *Differentiation*:

$$D^\alpha (f(x) \cdot g(y)) = D^\alpha f(x) \cdot g(y).$$

- *Multiplication*: if $a \in C^\infty(\mathbf{R}^n)$, then

$$a(x)(f(x) \cdot g(y)) = (a(x)f(x)) \cdot g(y).$$

- Let $x = (x_1, x_2, \ldots, x_n) \in \mathbf{R}^n$, $x^0 = (x_1^0, x_2^0, \ldots, x_n^0) \in \mathbf{R}^n$, then

$$\delta(x - x^0) = \delta(x_1 - x_1^0) \cdot \delta(x_2 - x_2^0) \cdot \ldots \cdot \delta(x_n - x_n^0). \tag{10.22}$$

The next properties we shall formulate as the following lemma:

Lemma 10.23. *Let $\tau(x, x^0)$ be a solution of* (10.1), (10.2); *and let $\Gamma = t^2 - \tau^2(x, x^0)$. Then under the assumptions* (A1) *and* (A2) *the following equalities hold:*

$$-\delta'(t) \cdot \theta(\Gamma) = 2\pi v^3(x^0)\delta(x - x^0) \cdot \delta(t) \begin{cases} 1, & k = -1, \\ 0, & k \geq 0, \end{cases} \tag{10.23}$$

$$-\delta(t) \cdot \theta(\Gamma) = 0, \tag{10.24}$$

$$\theta(t) \cdot \delta(\Gamma) = \frac{\delta(t - \tau(x, x^0))}{2\tau(x, x^0)}. \tag{10.25}$$

PROOF. Using the property (10.21) we have $-t\delta'(t) = \delta(t)$. Hence

$$-\delta'(t) \cdot \theta_k(\Gamma) = \frac{\delta(t)}{t} \cdot \theta_k(\Gamma).$$

The generalized function

$$\frac{\delta(t)}{t} \cdot \theta_k(\Gamma)$$

can be defined as the linear continuous functional on the space of test functions $D(\mathbf{R}^4)$ by means of the following relation:

$$\left(\frac{\delta(t)}{t} \cdot \theta_k(\Gamma)\phi(x, t)\right) = \int_{\mathbf{R}^4} \frac{\delta(t)}{t} \cdot \theta_k(\Gamma)\phi(x, t) \, dx \, dt$$

$$= \lim_{t \to +0} \frac{1}{t} \int_{\mathbf{R}^3} \theta_k(\Gamma)\phi(x, t) \, dx.$$

For the calculation of the last integral we introduce the curvilinear coordinates τ, θ, β of the point x using the equalities

$$x = f(\zeta, x^0), \qquad \zeta = \zeta' W(x^0), \qquad \zeta' = \tau\,(\sin\theta\cos\beta, \sin\theta\sin\beta, \cos\theta),$$

where $W(x^0) = (1/v(x^0))E$, E is the unit matrix. In this case, the curvilinear coordinate τ coincides with $\tau(x, x^0)$. Note that the following equalities hold:

$$\lim_{t \to +0} \frac{1}{t} \int_{\mathbf{R}^3} \theta_k(\Gamma)\phi(x, t) \, dx = \lim_{t \to +0} \frac{1}{t} \int_0^{2\pi} \int_0^{\pi} \int_0^{\infty} (t + \tau)^k \theta_k(t - \tau)\phi(x, t)$$

$$\times \left| \frac{\partial f(\zeta, x^0)}{\partial \zeta} \right| \Big|_{\zeta = \zeta' W(x^0)} \left| W(x^0) \right| \tau^2 \sin\theta \, d\tau \, d\theta \, d\beta$$

$$= 2\pi v^3(x^0)\phi(x^0, 0) \begin{cases} 1, & k = -1, \\ 0, & k \geq 0, \end{cases}$$

$$= (2\pi v^3(x^0)\delta(x - x^0, t), \phi(x, t)) \begin{cases} 1, & k = -1, \\ 0, & k \geq 0. \end{cases}$$

Hence

$$-\delta'(t) \cdot \theta_k(\Gamma) = 2\pi v^3(x^0)\delta(x - x^0, t) \begin{cases} 1, & k = -1, \\ 0, & k \geq 0, \end{cases}$$

and the property (10.23) is proved.

To prove (10.24) we use properties (10.18), (10.21), (10.22), (10.23). It is easy to see that the following equalities hold:

$$\delta(t) \cdot \theta_k(\Gamma) = -t\delta'(t) \cdot \theta_k(\Gamma) = 2\pi v^3(x^0)t\delta(x - x^0, t) \begin{cases} 1, & k = -1, \\ 0, & k \geq 0, \end{cases}$$

$$= -t\delta(t) \cdot 2\pi v^3(x^0)\delta(x - x^0) \begin{cases} 1, & k = -1, \\ 0, & k \geq 0, \end{cases} = 0.$$

The property (10.25) follows from the notion of the direct product and properties (10.17), (10.18).

10.4   The Fundamental Solution of the Cauchy Problem for the Acoustic Equation

In this section the notions of the generalized Cauchy problem and the fundamental solution of this problem are introduced. The existence theorem of the fundamental solution is proved here.

Consider the acoustic equation (10.3) subject to the condition (10.4). Let all coefficients of the acoustic equation be given infinitely differentiable functions of their arguments and we are looking for a generalized function $u(x, t) \in D'(\mathbf{R}^4)$ which satisfies (10.3) and (10.4). This problem is said to be the generalized Cauchy problem for the acoustic equation.

Definition 10.24. If the free term of the acoustic equation (10.3) is the Dirac delta function $\delta(x - x^0, t)$, then the solution of the generalized Cauchy problem is called the fundamental solution of the Cauchy problem for the acoustic equation.

The following existence theorem about the fundamental solution of the Cauchy problem for the acoustic equation takes place:

Theorem 10.25. *Under the assumptions* A1 *and* A2 *there exists the fundamental solution of the Cauchy problem* (10.3), (10.4) *having the form*

$$u(x, t, x^0) = \theta(t) \sum_{k=-1}^{\infty} \alpha_k(x, t, x^0)\theta_k(\Gamma), \qquad (10.26)$$

where $\theta_{-1}(\Gamma)$ is the Dirac delta function, $\theta_0(\Gamma) = \theta(\Gamma)$ is the Heaviside function

$$\theta_k(\Gamma) = \frac{t^k}{k!}\theta(\Gamma), \qquad k \geq 1, \quad \Gamma = t^2 - \tau^2(x, x^0),$$

and $\tau^2(x, x^0)$ is a solution of (10.1), (10.2):

$$\alpha_{-1}(x, x^0, t) = \frac{1}{2\pi v^3(x^0)} \left| \frac{\partial g(x, x^0)}{\partial x} \right|^{1/2} \frac{v(x)}{v(x^0)} \sqrt{\frac{m(x)}{m(x^0)}}$$

$$\times \exp\left[\frac{t}{2\tau(x, x^0)} \int_{T(x^0, x)} \sigma(\xi) d\tau \right], \qquad (10.27)$$

$$\alpha_0(x, t, x^0) = -\frac{\alpha_{-1}(x, t, x^0)}{4\tau(x, x^0)}$$

$$\times \int_{T(x^0, x)} [\alpha_{-1}(\xi, z, x^0)]^{-1} L_{\xi, z}\alpha_{-1}(\xi, z, x^0)\big|_{z=(t\tau(\xi, x^0))/\tau(x, x^0)} d\tau, \qquad (10.28)$$

$$\alpha_k(x, t, x^0) = -\frac{\alpha_{-1}(x, t, x^0)}{4}$$

$$\times \int_0^1 s^k [\alpha_{-1}(\xi, st, x^0)]^{-1} L_{\xi, z}\alpha_{-1}(\xi, z, x^0)\big|_{z=ts, \xi=f(g(x, x^0)s, x^0)} ds, \qquad (10.29)$$

where $T(x^0, x)$ is a ray connecting two points x^0 and x, $d\tau$ is an element of length of $T(x^0, x)$; and $|\partial g(x, x^0)/\partial x| > 0$ is the Jacobian of the transition from the

Cartesian to Riemann coordinate system

$$L_{\xi,z}\alpha_{-1}(\xi, z, x^0) = \frac{\partial^2}{\partial z^2}\alpha_{-1}(\xi, z, x^0) - v^2(\xi)\Delta\alpha_{-1}(\xi, z, x^0)$$

$$+ v^2(\xi)\nabla_\xi \ln m(\xi)\nabla_\xi\alpha_{-1}(\xi, z, x^0) - \sigma(\xi)\frac{\partial}{\partial z}\alpha_{-1}(\xi, z, x^0). \tag{10.30}$$

PROOF.   The fundamental solution of the Cauchy problem (10.3), (10.4) we seek in the form of the series expansion (10.26) in which $\alpha_k(x, t, x^0)$ are unknown functions. We have to determine these functions. For this aim we collect together all expressions by the functions $\theta_k(\Gamma)$. We use here the operations and properties of generalized functions, the eikonal equation (10.1), equalities

$$\frac{\partial\Gamma}{\partial t} = 2t, \qquad \nabla_x\Gamma = -2\tau(x, x^0)\nabla_x\tau(x, x^0),$$

$$\left(\frac{\partial\Gamma}{\partial t}\right)^2 - v^2(x)|\nabla_x\Gamma|^2 = 4\Gamma,$$

$$\Gamma\theta_{k-2}(\Gamma) = (k-1)\theta_{k-1}(\Gamma), \qquad \theta_k'(\Gamma) = \theta_{k-1}(\Gamma),$$

and the following expressions for u_t, u_{tt}, $\nabla_x u$, $\Delta_x u$:

$$\frac{\partial u}{\partial t} = \delta(t)\sum_{k=-1}^{\infty}\alpha_k\,\theta_k(\Gamma) + \theta(t)\frac{\partial}{\partial t}\sum_{k=-1}^{\infty}\alpha_k\,\theta_k(\Gamma)$$

$$= \theta(t)\sum_{k=-1}^{\infty}\left[\frac{\partial\alpha_{k-1}}{\partial t} + \frac{\partial\Gamma}{\partial t}\alpha_k\right]\theta_{k-1}(\Gamma),$$

$$\frac{\partial^2 u}{\partial t^2} = -\delta'(t)\sum_{k=-1}^{\infty}\alpha_k\theta_k(\Gamma) + 2\frac{\partial}{\partial t}\left[\delta(t)\sum_{k=-1}^{\infty}\alpha_k\,\theta_k(\Gamma)\right]$$

$$+ \theta(t)\frac{\partial^2}{\partial t^2}\sum_{k=-1}^{\infty}\alpha_k\theta_k(\Gamma) = 2\pi v^3(x^0)\alpha_1(x^0, 0, x^0)\delta(x - x^0, t)$$

$$+ \theta(t)\sum_{k=-1}^{\infty}\left[\frac{\partial^2\alpha_{k-1}}{\partial t^2} + 2\frac{\partial\alpha_k}{\partial t}\frac{\partial\Gamma}{\partial t} + \alpha_k\frac{\partial^2\Gamma}{\partial t^2} + \left(\frac{\partial\Gamma}{\partial t}\right)^2\frac{(k-1)}{\Gamma}\alpha_k\right]\theta_{k-1}(\Gamma),$$

$$\nabla_x u = \theta(t)\sum_{k=-1}^{\infty}\left[\nabla_x\alpha_{k-1} + \alpha_k\nabla_x\Gamma\right]\theta_{k-1}(\Gamma),$$

$$\Delta_x u = \theta(t)\sum_{k=-1}^{\infty}\left[\Delta_x\alpha_{k-1} + 2\nabla_x\alpha_k\nabla_x\Gamma + \alpha_k\Delta_x\Gamma\right.$$

$$\left. + \alpha_k(\nabla_x\Gamma)^2\frac{(k-1)}{\Gamma}\right]\theta_{k-1}(\Gamma),$$

where $\alpha_2 \equiv 0$.

After substitution (10.26) into (10.3) we get

$$\left[2\pi v^3(x^0)\alpha_{-1}(x^0,0,x^0)) - 1\right]\delta(x-x^0,t)$$

$$+ \theta_0(t)\sum_{k=-1}^{\infty}\left[L_{x,t}\alpha_{k-1} + 2\frac{\partial\alpha_k}{\partial t}\frac{\partial\Gamma}{\partial t} - 2v^2(x)\nabla_x\alpha_k\nabla_x\Gamma\right.$$

$$+ \left(L_{x,t}\Gamma + 4(k-1)\right)\alpha_k\Big]\theta_{k-1}(\Gamma) = 0. \tag{10.31}$$

Equating to zero the expressions by $\delta(x-x^0,t)$ and $\theta_{k-1}(\Gamma), k = -1,0,1,2,\ldots,$ and using the formula

$$L_{x,t}\Gamma \equiv \frac{\partial^2}{\partial t^2}\Gamma - v^2(x)\Delta_x\Gamma + v^2(x)\nabla_x\ln m(x)\nabla_x\Gamma - \sigma(x)\frac{\partial}{\partial t}\Gamma$$

$$= 2 - \text{div}_x\left(v^2(x)\nabla_x\Gamma\right) - 2\tau(x,x^0)\nabla_x v^2(x)\nabla_x\tau(x,x^0)$$

$$- 2\tau(x,x^0)v^2(x)\nabla_x\ln m(x)\nabla_x\tau(x,x^0) - 2t\sigma(x),$$

we have the relations

$$\alpha_{-1}(x^0,0,x^0) = \frac{1}{2\pi v^3(x^0)},$$

$$4t\frac{\partial\alpha_k}{\partial t} + 4\tau(x,x^0)v^2(x)\nabla_x\alpha_k\nabla_x\tau^2(x,x^0) + \left(2 + 4(k-1)\right)$$

$$- \text{div}_x\left(v^2(x)\nabla_x\Gamma)\right)\alpha_k - 2\tau(x,x^0)v^2(x)\left[\nabla_x\ln v^2(x)\right.$$

$$+ \nabla_x\ln m(x)\Big]\nabla_x\tau(x,x^0)\alpha_k - 2t\sigma\alpha_k + L_{x,t}\alpha^{k-1} = 0, \quad (10.32)$$

$$k = -1,0,1,2,\ldots, \qquad \alpha_{-2} \equiv 0.$$

Along the ray $T(x^0,x)$ the following equalities are valid

$$\frac{dx}{d\tau} = v^2(x)\nabla_x\tau(x,x^0), \qquad \frac{t}{\tau} = p.$$

Here p is a constant and τ is a parameter which we introduce by $\tau = \tau(x,x^0)$. Using Lemma 10.12 the differential relation (10.33) can be written along the curve $T(x^0,x)$ as follows:

$$4\tau\frac{\partial}{\partial\tau}\left[\alpha_k(x(\tau),p\tau,x^0)\right] + \left[4(k+1) - 2\tau\frac{\partial}{\partial\tau}\ln\left|\frac{\partial g(x,x^0)}{\partial x}\right| - 2\tau p\sigma(x)\right.$$

$$\left.-2\tau\frac{\partial}{\partial\tau}\left(\ln v^2(x)m(x)\right)\right]\Bigg|_{x=x(\tau)} \alpha_k(x(\tau),p\tau,x^0)$$

$$= L_{x,z}\alpha_{k-1}(x,z,x^0)\big|_{x=x(\tau),z=p\tau}, \qquad k = -1,0,1,2,\ldots, \qquad \alpha_{-2} \equiv 0.$$

Multiplying, by the factor

$$\tau^2\exp\left(-\frac{p}{2}\int_0^\tau \tau\sigma(x(\tau))\,d\tau\right),$$

both sides of the last relations, we can reduce them to the following:

$$\frac{\partial}{\partial\tau}\left[\Psi(x(\tau),p\tau,x^0)\alpha_{-1}(x(\tau),p\tau,x^0)\right], \qquad k = -1, \tag{10.33}$$

$$\frac{\partial}{\partial \tau}\left[\tau^{k+1}\Psi(x(\tau), p\tau, x^0)\alpha_k(x(\tau), p\tau, x^0)\right]$$

$$= -\frac{\tau^k}{4}\Psi(x(\tau), p\tau, x^0)L_{x,t}\alpha_{k-1}(x, t, x^0)\Big|_{x=x(\tau),t=p\tau}, \qquad k \geq 0. \quad (10.34)$$

Here

$$\Psi(x, t, x^0) = \left(v(x)\sqrt{m(x)\left|\frac{\partial g(x, x^0)}{\partial x}\right|}\right)^{-1}$$

$$\times \exp\left(-\frac{t}{2\tau(x, x^0)}\int_{T(x,x^0)}\sigma(\xi)\,d\tau\right). \quad (10.35)$$

The formula (10.27) immediately follows from (10.34), (10.32). Integrating (10.35) from 0 to $\tau(x, x^0)$ we find

$$\tau^{k+1}(x, x^0)\Psi(x, t, x^0)\alpha_k(x, t, x^0) = -\frac{1}{4}\int_0^{} \tau(x, x^0)\tau^k\Psi(\xi, z, x^0)$$

$$\times L_{\xi,z}\alpha_{k-1}(\xi, z, x^0)\Big|_{\xi=f(\zeta,x^0),z=t/(\tau(x,x^0)\tau)}d\tau, \quad (10.36)$$

where $\zeta = \alpha\tau(x, x^0)$ is the Riemann coordinates of the point x.

Making the change of variable τ on s by the rule $\tau = \tau(x, x^0)s$ we reduce the last integral to the form

$$-\frac{1}{4}\tau^{k+1}(x, x^0)\int_0^1 s^k\Psi(\xi, z, x^0)$$

$$\times L_{\xi,z}\alpha_{k-1}(\xi, z, x^0)\Big|_{\xi=f(g(x,x^0)s,x^0),z=ts}. \quad (10.37)$$

The formulas (10.28), (10.29) for an arbitrary $k \geq 0$ follow from (10.27), (10.36)–(10.38). $\qquad \square$

10.5 Properties of the Fundamental Solution

Let $u(x, t, x^0)$ be the solution of the problem (10.3), (10.4) having the form (10.26) and let $h(x, x^0)$, $f(x, x^0)$ be two functions defined by the formulas

$$h(x, x^0) = \lim_{t\to\tau(x,x^0)+0}\int_\infty^t u(x, z, x^0)\,dz, \quad (10.38)$$

$$f(x, x^0) = \lim_{t\to\tau(x,x^0)+0} u(x, z, x^0). \quad (10.39)$$

Theorem 10.26. *Under assumptions A1, A2 and the above-mentioned notations the following relations take place*

$$h(x, x^0) = \frac{v(x)}{4\pi\tau(x, x^0)v^4(x^0)}\sqrt{\frac{m(x)}{m(x^0)}}$$

$$\times \exp\left(-\frac{t}{2\tau(x, x^0)}\int_{T(x,x^0)}\sigma(\xi)\,d\tau\right). \quad (10.40)$$

$$\int_{T(x^0,x)} U(\xi)\,d\tau = F_2(x, x^0), \qquad x \in S, \quad x^0 \in S, \quad (10.41)$$

where

$$U(x) = \frac{v^2(x)}{2}\left[\Delta_x R(x) - \tfrac{1}{2}|\nabla_x R|^2\right], \qquad R(x) = \ln m(x), \quad (10.42)$$

$$F_2(x, x^0) = 2\frac{f(x, x^0)}{h(x, x^0)}$$

$$+ \int_{T(x,x^0)} B^2(x^0, \xi)\,d\tau_\xi - \int_{T(x,x^0)} \sigma(\xi)B(x^0, \xi)\,d\tau_\xi$$

$$- \int_{T(x,x^0)} v^2(\xi)\frac{\Delta_\xi A(\xi, t, x^0)}{A(\xi, t, x^0)}\bigg|_{t=\tau(\xi,x^0)}\,d\tau_\xi, \qquad (10.43)$$

$$A(x, t, x^0, t) = \frac{v(x)}{2\pi v^4(x^0)}\left|\frac{\partial g(x, x^0)}{\partial x}\right|^{1/2}\sqrt{\frac{1}{m(x^0)}}$$

$$\times \exp\left(\frac{t}{2\tau(x, x^0)}\int_{T(x^0,x)} \sigma(\xi)\,d\tau_\xi\right),$$

$$B(x^0, \xi) = \frac{1}{2\tau(\xi, x^0)}\int_{T(x^0,\xi)} \sigma(\zeta)\,d\tau_\zeta.$$

PROOF.   Using (10.26), (10.25) we have

$$h(x, x^0) = \frac{\alpha_{-1}(x, \tau(x, x^0)+0, x^0)}{2\tau(x, x^0)}. \qquad (10.44)$$

Formula (10.41) follows from the last one and (10.27).

  Consider now the formula (10.40). From (10.26), (10.28) it follows that

$$f(x, x^0) = \alpha_0(x, \tau(x, x^0)+0, x^0) = -\frac{\alpha_{-1}(x, \tau(x, x^0)+0, x^0)}{4\tau(x, x^0)}$$

$$- \int_{T(x^0,x)}\left[\frac{L_{\xi,z}\alpha_{-1}(\xi, z, x^0)}{\alpha_{-1}(\xi, z, x^0)}\right]\bigg|_{t=\tau(\xi,x^0)}\,d\tau_\xi. \qquad (10.45)$$

We can obtain that

$$\frac{L_{x,t}\alpha_{-1}(x, t, x^0)}{\alpha_{-1}(x, t, x^0)} = -\frac{v^2(x)}{2}\left[\Delta_x R(x) - \tfrac{1}{2}|\nabla_x R|^2\right]$$

$$+ B^2(x^0, x) - \sigma(x)B(x^0, x) - v^2(x)\frac{\Delta_x A(x, t, x^0)}{A(x, t, x^0)}, (10.46)$$

where functions $R(x)$, $B(x^0, x)$, and $A(x, t, x^0)$ were introduced in Theorem 10.26. Using (10.45)–(10.47) we get at first

$$2\frac{f(x, x^0)}{h(x, x^0)} = -\int_{T(x^0,x)}\frac{L_{\xi,z}\alpha_{-1}(\xi, z, x^0)}{\alpha_{-1}(\xi, z, x^0)}\bigg|_{t=\tau(\xi,x)}\,d\xi,$$

and then (10.42). The theorem is proved.   □

10.6   The Scheme of Solving the Inverse Dynamic Problem

Consider now the acoustic equation (10.3) subject to the condition (10.4). Let $D \subset \mathbf{R}^3$ be a fixed limited domain with the smooth boundary S. Moreover, $D \subset \mathbf{R}^3$ is convex with respect to the family of rays $\{T(x^0, x)\}, x \in D, x^0 \in D$. This means that for any pair of points x^0, x from D the ray $T(x^0, x)$ belongs to D. We assume that the infinite differentiable functions $v(x), m(x), \sigma(x)$ have constant values v^0, m^0, σ^0 for $x \in \mathbf{R}^3 \backslash \bar{D}$ and the values of these functions are unknown for $x \in D$. We consider the problem of finding these unknown functions if the information of the form (10.5) is given. We note that the function $G(x, t, x^0)$ appearing in (10.5) is known for $x \in S, x^0 \in S, t \in [0, T]$; T is here a sufficiently large number. Note also that the information (10.5) contains the data about $\tau(x, x^0), x \in S, x^0 \in S$. Thus it leads to the inverse kinematic problem of finding $v(x), x \in D$. This inverse kinematic problem consists in determining, in D, the function $v(x)$ which enters into the eikonal equation (10.1) if we know some information about the solution satisfying the data (10.2). Namely, this information is of the form

$$\tau(x, x^0)\big|_{x \in S, x^0 \in S} = H(x, x^0), \tag{10.47}$$

where $H(x, x^0)$ is a given function for $x \in S, x^0 \in S$.

The solution of this inverse kinematic problem is the first step in solving the inverse dynamic problem. We will not develop the description of this problem solution although this is an interesting topic. We note that in [AL1], [AL2], [AN], numerical algorithms are given for solving the multidimensional inverse kinematic problem. Further, we assume that the inverse kinematic problem can be solved. Therefore, if we know the function $v(x)$ for $x \in \mathbf{R}^3$, then we find the family of the rays $T(x, x^0), x \in S, x^0 \in S$, and values of the function $\tau(x, x^0), x \in \mathbf{R}^3, x^0 \in \mathbf{R}^3$. Using the supplementary data (10.5) we can find the values of the function $h(x, x^0)$ for $x \in S, x^0 \in S$ by the formula

$$h(x, x^0) = \lim_{t \to \tau(x, x^0) + 0} \int_\infty^t G(x, z, x^0)\, dz.$$

Noting that $v(x) \equiv v^0, m(x) \equiv m^0$ for $x \in S$, we obtain from (10.41) the formula

$$\int_{T(x^0, x)} \sigma(\xi)\, d\tau = F_1(x, x^0), \qquad x \in S, \quad x^0 \in S, \tag{10.48}$$

where

$$F_1(x, x^0) = \ln \left(4\pi \tau(x, x^0) v(x^0) h(x, x^0) \right)^2 \left| \frac{\partial g(x, x^0)}{\partial x} \right|^{-1}.$$

The right-hand side of (10.48) is a known function for $x \in S, x^0 \in S$.

The second step of solving the inverse problem is connected with the solution of the integral geometry problem (10.48). The last one consists in finding the unknown function $\sigma(x)$ in the domain D if we know the values of the function

$F_1(x, x^0)$ for all $x \in S$, $x^0 \in S$, i.e., all integrals of the unknown function $\sigma(x)$ are given along each ray $T(x, x^0)$ which connects arbitrary pair of points x^0, x from S. Such a kind of the integral geometry problems was first solved by Radon [RA]. It was then investigated in various aspects by John [J1], [J2], Khachaturov [KH], Kostelyanets and Reshetnyak [KR], Gel'fand and Graev [GG], [GGV], and others. The theory of the integral geometry problems was further developed in the works of Lavrent'ev and Bukhgeim [LB1], [LB2], [BU], Romanov [ROM2], Mukhametov [MU], Natterer [NA], and other authors.

After finding the functions $v(x)$ and $\sigma(x)$, the values of the function $F_2(x, x^0)$ can be defined for $x \in S$, $x^0 \in S$ by formula (10.44).

The function $U(x)$ can be found by means of solving the integral geometry problem (10.42) which is the next, third step, of solving our inverse problem.

The function $R(x)$ is determined as the solution of the Dirichlet problem in the domain D for the elliptic equation (10.43) subject to the Dirichlet condition

$$R|_S = \ln m^0.$$

Here $v(x)$ and $U(x)$ are known functions.

Then function $m(x)$ is defined by formula

$$m(x) = \exp(R(x)),$$

and thus the functions $v(x)$, $m(x)$, $\sigma(x)$ are found. The multidimensional inverse problem for the acoustic equation is solved.

10.7   Remarks

Some inverse problems in the ray statement for the Lamé system of elasticity and Maxwell's system of electrodynamics were studied in the works [YA1], [YA2]. From a mathematical point of view the present work continues the investigations of [ROM1], [YA1], [ROM3], [YA2].

Let us briefly describe the results of the investigations of the multidimensional inverse problem in ray statements for the acoustic equation with the constant velocity and the case when all function-coefficients have limited smoothness.

Remark 10.27 (Reduction of the Inverse Problem in the Ray Statement with a Constant Velocity to Tomography Problems).   Let now $v(x) \equiv v^0$ for $x \in \mathbf{R}^3$, v^0 be a constant,

$$D = \{x \in \mathbf{R}^3 \mid |x| < a\}, \qquad S = \{x \in \mathbf{R}^3 \mid |x| = a\},$$
$$m(x), \sigma(x) \in C^\infty, \quad m(x) > 0, \quad \sigma \equiv \sigma^0, \quad m \equiv m^0 \quad \text{for} \quad x \in \mathbf{R}^3 \backslash D,$$
$$m^0, \quad \sigma^0, \quad \text{be constants}, \quad m^0 > 0.$$

Using the above-described reasoning we can show that in this case the inverse problem can be reduced to the following two tomography, and one Dirichlet's, problems. The first tomography problem is

$$\int_{L(x^0,x)} \tilde{\sigma}(\xi)ds = F_1(x, x^0), \quad x \in S, \quad x^0 \in S, \tag{10.49}$$

where

$$F_1(x, x^0) = \ln\left(4\pi \left|x - x^0\right| (v^0)^2 h(x, x^0)\right)^2 - \int_{L(x^0,x)} \sigma^0\, ds,$$

where $L(x, x^0)$ is the intercept of the straight line connecting the points x^0 and x; and ds is the element of length of the straight line. This tomography problem consists in determining the unknown function $\tilde{\sigma}(x)$ inside D if we know all the integrals of the function $\tilde{\sigma}(x)$ along the intercepts of the straight lines which connect arbitrary points x^0 and x from S.

The second tomography problem is to determine the function $U(x)$ inside D if we know

$$\int_{L(x^0,x)} U(\xi)\, ds = F_2(x^0, x), \tag{10.50}$$

where $F_2(x, x^0)$ is the known function with values found by formulas (10.44). In this case $T(x^0, x)$ is the intercept of the straight line $L(x^0, x)$, and the following correlations hold:

$$\tau(x, x^0) = \frac{\left|x - x^0\right|}{v^0},$$

$$A(x, t, x^0) = \frac{1}{2\pi v^3(x^0)}\sqrt{\frac{1}{m(x^0)}}\, \exp\left(\frac{t}{2}\int_0^1 \sigma(x^0 + s(x - x^0))ds\right).$$

If we know the function $U(x)$, then we obtain the Dirichlet problem of determining the function $R(x)$ inside D for which the following equalities take place:

$$\Delta_x R(x) - \tfrac{1}{2}|\nabla_x R|^2 = \frac{2}{(v^0)^2}U(x), \quad x \in D,$$

$$R\big|_{x \in S} = \ln m_0.$$

Having solved this Dirichlet problem we can find the function $m(x)$ by

$$m(x) = \exp(R(x)).$$

The tomography problems (10.49) and (10.50) can be rewritten using the notation of [MR] as

$$(\mathcal{P}\tilde{\sigma})(y, \omega) = f_j(y, \omega), \quad j = 1, 2,$$

$$y \in \mathbf{R}^3, \quad \omega \in S^2 = \left\{\omega \in \mathbf{R}^3 \mid |\omega| = 1\right\},$$

where

$$(\mathcal{P}\tilde{\sigma})(y, \omega) \equiv \int_{-\infty}^{+\infty} \tilde{\sigma}(y + s\omega)\, ds.$$

and where the functions $f_j(y, \omega)$, $j = 1, 2$, are defined by the following rule. For a fixed point $(y, \omega) \in \mathbf{R}^3 \times S^2$ we consider the straight line $\bar{x} = y + \omega s$, $s \in \mathbf{R}$. If the straight line $\bar{x} = y + \omega s$ intersects S at the points x^0, x, then the value of the function $f_j(y, \omega)$ is found by $f_j(y, \omega) = F_j(x, x^0)$, and if the straight line $\bar{x} = y + \omega s$ does not intersect $\overline{D}$, then $f_j(y, \omega) = 0$, $j = 1, 2$.

Since rays became the straight lines for the constant velocity then the integral geometry problem is transformed to a tomography one. We note that there are uniqueness, stability estimates, existence theorems, and numerical methods for solving tomography problems which can be found in Nattere's book [NA].

Remark 10.28 (The Structure of the Fundamental Solution of the Cauchy Problem and the Solution of the Inverse Dynamic Problem for the Limited Smoothness of Function-Coefficients). The assumption about infinite differentiation of the coefficients of the acoustic equation can be changed on the following conditions:

$$v(x) \in C^5(\mathbf{R}^2), \qquad \sigma(x) \in C^4(\mathbf{R}^2), \qquad m(x) \in C^4(\mathbf{R}^2).$$

In this case we can show that the fundamenal solution has the form

$$u(x, x^0, t) = \theta(t) \left[\alpha_{-1}(x, t, x^0) \delta(\Gamma) + \alpha_0(x, t, x^0) \theta(\Gamma) \right] + \tilde{u}(x, t, x^0), \quad (10.51)$$

where $\tilde{u}(x, t, x^0)$ is a continuous function having the structure

$$\tilde{u}(x, t, x^0) = \left(t^2 - \tau^2(x, x^0) \right) \theta(t^2 - \tau^2(x, x^0)) V(x, t, x^0),$$

here $V(x, t, x^0)$ is a limited function in every closed domain and the functions $\alpha_1(x, t, x^0), \alpha_0(x, t, x^0)$ are defined by (10.27), (10.28). All reasonings of Sections 10.5 and 10.6 are the same for this case. Hence the inverse dynamic problem for the acoustic equation has the same scheme and may be solved.

We note also that there are the uniqueness and stability estimate theorems for the inverse kinematic problem with the function velocity having bounded smoothness and satisfying (A1), (A2) (see [MR], [ROM1]). The uniqueness and stability estimate theorems of the integral geometry problem take place with these restrictions on the velocity function (see [BG], [ROM1], [ROM2], [MU]). Summarizing these results we can obtain the uniqueness and stability estimate theorems for our original dynamic inverse problem for the acoustic equation. Some numerical methods of the solution of the inverse kinematic and integral geometry problems can be found in [AL1], [AL2], [AN]. Developing new efficient numerical algorithms for the inverse kinematic and integral geometry problems is a very actual topic now.

Acknowledgments. I am grateful to Professor Vladimir Romanov for bringing me into the topic of the multidimensional inverse problems in ray statements for hyperbolic equations. Vladimir Romanov suggested that I investigate the multidimensional inverse problem for Lame's system of elasticity several years ago.

The faculty of Arts and Sciences of the Dokuz Eylul University in Izmir provided me with the harmonious environment in the period of writing this chapter. My special thanks to Professor Guzin Gokmen, Professor Gonga Onargan and to my wife Professor Tatyana Yakhno, for all their attention and support during my work.

10.8   References

[AL1]   A.S. Alekseev. A numerical method for solving the three-dimensional inverse kinematic problem of seismology. *Mat. Problemy Geofiz.*, 1:176–201, 1969 (Russian).

[AL2]   A.S. Alekseev. A numerical method for determining the structure of the Earth's upper mantle. *Mat. Problemy Geofiz.*, 2:141–165, 1971 (Russian).

[ANP]   Yu.E. Anikonov, N.B. Pivovarova, and L.V. Slavina. The three-dimensional velocity field of the Kamchatka focal zone. *Mat. Problemy Geofiz.*, 5:92–117, 1974 (Russian).

[AN]   Yu.E. Anikonov. *Some Methods for the Study of Multidimensional Inverse Problems for Differential Equations*. Nauka, Novosibirsk, 1978 (Russian).

[BE]   G.Ya. Beil'kin. Stability and uniqueness of the solution of the inverse kinematic problem of seismology in higher dimensions. *Soviet Math. Dokl.*, 21:3–6, 1983.

[BG]   I.N. Bernshtein and M.L. Gerver. A problem of integral geometry for a family of geodesics and an inverse kinematic problem of seismology. *Dokl. Earth Sci. Sections*, 243:302–305, 1978.

[BU]   A.L. Bukhgeim. Some problems of integral geometry. *Siberian Math. J.*, 13:34–42, 1972.

[EL]   L.E. El'sgol'tz. *Differential Equations and Calculus of Variations*. Mir, Moscow, 1970.

[GG]   I.M. Gel'fand and M.I. Graev. *Geometry of Homogeneous Spaces, Representations of Groups in Homogeneous Spaces and Related Questions of Integral Geometry*. American Mathematical Society Translations, Providence, RI, 1964.

[GGV]   I.M. Gel'fand , M.I. Graev, and N.Ya. Vilenkin. *Generalized Functions, Integral Geometry and Representation Theory*. Academic Press, London, 1966.

[GM]   M.L. Gerver and V.M. Markushevich. Determining seismic-wave velocities from travel-time curves. *Comput. Seism.*, pp. 3–51, Plenum Press, New York, 1972.

[HR]   G. Herglotz. Uber die elastizitat the erde bei berucksichtigung ihrer variablen dichte. *Z. Math. Phys.*, 52:275–299, 1905.

[J1]   F. John. Bestimmung einer funktion aus ihren integralen uber gewisse manningfaltigkeiten. *Math. Ann.*, **109**:488–520, 1933.

[J2]   F. John. Abhangigkeiten zwishen den flachen integralen einer stetingen funktionen. *Math. Ann.*, **111**:541–559, 1935.

[KH]   A.A. Khachaturov. Determination of the value of the measure of region of n-dimensional Euclidean space from its values for all half-spaces. *Uspekhi Mat. Nauk*, **9**:205–212, 1954 (Russian).

[KR]   P.O. Kostelyanetz and Yu.G. Reshetnyak. Determination of a completely additive function by its values on half-spaces. *Uspekhi Mat. Nauk*, **9**:131–140, 1954 (Russian).

[LB1]   M.M. Lavrent'ev and A.L. Bukhgeim. On a class of problems of integral geometry. *Soviet Math. Dokl.*, **14**:38–39, 1973.

[LB2]   M.M. Lavrent'ev and A.L. Bukhgeim. On a class of operator equations of the first kind. *Functional Anal. Appl.*, **7**:44–53, 1973.

[LRS]   M.M. Lavrent'ev, V.G. Romanov, and S.P. Shishatskii. *Ill-Posed Problems of Mathematical Physics and Analysis*. American Mathematical Society Translations, Providence, RI, 1986.

[MU]   R.G. Mukhometov. The problem of reconstructing a two-dimensional Riemannian metric and integral geometry. *Soviet Math. Dokl.*, **18**:32–35, 1977.

[MR]   R.G. Mukhametov and V.G. Romanov. On the problem of finding an isotropic Riemannian metric in n-dimensional space. *Soviet Math. Dokl.*, **19**:1279–1281, 1978.

[NA]   F. Natterer. *The Mathematics of Computerized Tomography*. Wiley, Stuttgart, and B.G. Tenbner, Leipziz, 1986.

[RA]   J. Radon. Uber die bestimmungen von funktionen durch ihre integralwerte langs gewisser manningfaltigkeiten. *Ber. Verh. Sachs. Ges. Wiss. Leipzig Math.-Phys. Kl.*, 262–277, 1917.

[ROM1]   V.G. Romanov. *Inverse Problems of Mathematical Physics*. VNU Science Press, Utrecht, The Netherlands, 1987.

[ROM2]   V.G. Romanov. Integral geometry on the geodesics on isotropic Riemannian metric. *Soviet Math. Dokl.*, **19**:290–293, 1978.

[ROM3]   V.G. Romanov. Structure of the fundamental solution of the Cauchy problem for Maxwell's systems of equations. *Differents. Uravn.*, **22**:1577–1587, 1986 (Russian).

[ROMK]   V.G. Romanov and S.I. Kabanikhin. *Inverse Problems for Maxwell's Equations*. VNU Science Press, Utrecht, The Netherlands, 1994.

[VL]   V.S. Vladimirov. *Equations of Mathematical Physics*. Marcel Dekker, New York, 1971.

[YA1]   V.G. Yakhno. *Inverse Problems for Differential Equations of Elasticity*. Nauka, Novosibirsk, 1990 (Russian).

[YA2]   V.G. Yakhno. Multidimensional inverse problems in the ray formulation for hyperbolic equations. *J. Inverse Ill-Posed Problems*, 6:373–386, 1998.

[WK]   E. Wiechert, K. Zoeppritz. Uber erdbebenwellen. *Nachr. Konigl. Ges. Wiss. Göttingen Math.-Phys. Kl.*, 4:415–549, 1907.

APPENDIX

Ocean Acoustic Tomography: Integral Data and Ocean Models[1]

Bruce D. Cornuelle
Peter F. Worcester

ABSTRACT Tomographic data differ from most other oceanographic data because their sampling and information content are localized better in spectral space than in physical space. Approximate data assimilation methods optimized for localized physical space measurements can lose much of the nonlocal tomographic information, degrading the potential performance of the tomographic data in fixing the model state. In addition, methods which require inverting the tomographic data to a physical space grid before insertion into the model as point measurements incur special problems because of the nonlocal nature of the errors. We give a simple one-dimensional example to illustrate the problems that can arise. Fortunately, methods that directly assimilate integral measurements and preserve all of the information in the tomographic data, such as the Kalman filter, are becoming more practical as computer speed grows.

A.1 Introduction

Ocean acoustic tomography is a remote sensing technique that exploits the transparency of the ocean to low-frequency sound and the sensitivity of acoustic propagation to the ocean sound speed (temperature) and current fields (Munk and Wunsch, 1979). The travel times of acoustic pulses or the group delays of acoustic normal modes are interpreted to provide information about the intervening ocean. Tomographic data differ from more familiar data types, such as moored temperature and current measurements or hydrographic data, in that the data are integrals of the temperature and current fields along the acoustic paths, rather than values at a single location.

Sound speed perturbations are at least an order of magnitude greater than current speeds, dominating travel time perturbations. Measurements of travel times between multiple points in the ocean can therefore be used to estimate the ocean sound speed field. The effects of sound speed perturbations and currents can be separated using acoustic transceivers, rather than separate sources and receivers. The difference in travel times of acoustic pulses traveling in opposite directions is to

[1]Reprinted from P. Melanotte-Rizzoli (editor), Modern Approaches to Data Assimilation in Ocean Modelling, 1996, pp. 97–115, with permission from Elsevier Science.

first order proportional only to the integral of current along the path. The sum of the travel times is proportional only to the integral of the sound speed field. Reciprocal transmissions can therefore be used to measure both sound speed and current.

Linear inverse methods are commonly used to estimate the sound speed and current fields from tomographic data. These methods are well developed for data collected more or less simultaneously, but much less effort has been devoted to the methods needed to combine data collected at different times. Howe et al. (1987) applied a Kalman filter to combine data collected at different times from a single source–receiver pair, with the assumption that the ocean tends to climatology with a 10-day time constant (i.e., assuming that the perturbation state vector decays to zero with a 10-day time constant). Spiesberger and Metzger (1991) employed a generalized objective mapping approach in which a temporal covariance function was specified, in addition to the usual spatial covariance function, to analyze travel-time time series from a different experiment.

Neither of these approaches takes advantage of our knowledge of ocean dynamics to constrain how the temperature and current fields evolve. Chiu and Desaubies (1987) included dynamics in the analysis of data from the 1981 Tomography Demonstration Experiment (Ocean Tomography Group, 1982; Cornuelle et al., 1985) in perhaps the simplest possible way, assuming that the ocean perturbations were made up of linear Rossby waves and using the data to estimate time-independent amplitudes and phases of selected Rossby wave components in a global (nonlinear) least squares fit. Cornuelle (1990) used a a Kalman filter with a linear ocean model to combine data obtained at different times in simulated moving ship tomography experiment. More generally, however, fully nonlinear ocean general circulation models (GCMs) are required to describe realistically the evolution of the ocean. Schröter and Wunsch (1986) described one approach to forcing a steady ocean GCM to consistency with a comparatively sparse set of tomographic measurements, but did not consider the time-dependent problem. Munk et al. (1995) and Wunsch (1990) discuss the general problem of combining tomographic data with time-dependent ocean models to obtain the best possible estimate of the state of the ocean. Sheinbaum (1989) performed simulations in which tomographic measurements were combined with a simple, time-dependent, one-layer ocean model. Fukumori and Malanotte–Rizzoli (1995) constructed an approximate Kalman filter for a nonlinear, primitive equation model of the Gulf Stream, and examined the assimilation of various pseudomeasurements, including tomographic observations. The same fundamental methods used for other oceanographic data are in principal also applicable to tomographic data.

There are practical difficulties due to the large number of state variables in ocean GCMs, however. The computer time and memory required to implement exact assimilation methods have led to the development of various approximate methods for assimilating point measurements into ocean GCMs. Unfortunately, approximate methods suitable for point measurements often are not suitable for tomographic data, due to the quite different ways in which point sensors and tomographic measurements sample the ocean. Sequential methods that do not retain the full uncertainty covariance, for example, lose significant information from step to step when used with tomographic data. In this Appendix, it is not our goal to review the inverse methods required to combine tomographic measurements with ocean models, for which we refer the reader to Munk et al. (1995). Rather, we review the sampling properties of tomographic measurements, and present a cautionary note as to what can go wrong when approximate methods developed for point measurements are blindly applied to tomographic data.

A.2   Sampling Properties of Acoustic Rays

Most tomographic applications outside the earth sciences are characterized using spectral methods, such as the projection-slice theorem central to the reconstruction in medical tomography. The irregularity of the ray paths in geophysical and ocean tomography destroys the simplest spectral relationships, but the nonlocal character of the data remains.

A.2.1   Vertical Slice: Range-Independent

A measurement of acoustic ray travel time is sensitive to the integral of sound speed and tangential current along the entire ray path, although the weighting is not uniform in range, because ray travel time is most sensitive to propagation speed perturbations at the upper and lower ray turning points. This depth dependence can be seen most easily in the range-independent case (Munk and Wunsch, 1982; Cornuelle et al., 1993). To first order, ray travel time perturbations are weighted averages of the sound-speed perturbations integrated along the unperturbed ray paths, Γ_i:

$$\Delta T_i = - \int_{\Gamma_i} \frac{ds}{C_0^2(x, z)} \Delta C(x, z),$$

where ocean currents have been neglected. For the range-independent case, in which $C_0(z)$ and $\Delta C(z)$ are independent of x, this can be converted to an integral over z:

$$\Delta T_i = - \int_{\Gamma_i} \frac{dz}{C_0^2 \sqrt{1 - (C_0/\hat{C})^2}} \Delta C(z),$$

sing Snell's law, $C(z)/\cos\theta = \hat{C}$. $\hat{C}$ is the sound speed at the ray turning point and θ is the angle relative to horizontal. The function

$$\frac{1}{C_0^2(z)\sqrt{1 - (C_0(z)/\hat{C})^2}}$$

gives the weighting with which the $\Delta C(z)$ contribute to ΔT_i. The (integrable) singularity at the ray turning point depths, $\hat{z}$, where $C_0(\hat{z}) = \hat{C}$, clearly shows that ray travel times are most sensitive to the ocean at ray turning point depths. There is an ambiguity in that singularities occur at both the upper and lower turning points.

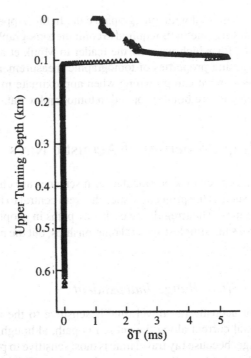

FIGURE A.1. Travel time anomalies computed for a sound speed perturbation with an amplitude of −1 m/s at 100 m depth linearly decreasing to zero at 90 and 110 m. The unperturbed profile is range-independent. (Reproduced from Cornuelle et al., 1993.)

As an example, Cornuelle et al. (1993) computed the travel time perturbations due to a sound speed perturbation with an amplitude of −1 m/s at 100 m depth, linearly decreasing to zero at 90 and 110 m. The travel time anomaly, computed by subtracting the travel time in the unperturbed ocean from the travel time in the perturbed ocean, is sharply peaked for rays that turn between 90 and 110 m (Figure A.1). The anomaly is zero for rays with upper turning depths below 110 m because they do not sample the perturbed region. Rays that have upper turning depths above 90 m have nonzero anomalies, because they traverse the perturbed region, but the anomalies are relatively small because the ray weighting function falls off rapidly with distance from the turning point.

A.2.2   Vertical Slice: Range-Dependent

A ray trapped in the sound channel, with upper and lower turning points at regular intervals, samples the ocean periodically in space, so that its travel time is sensitive to some spatial frequencies, but unaffected by others. Chiu et al. (1987) found from simulations, and Howe et al. (1987) showed experimentally, that some range-dependent information on the sound speed field in a vertical slice can be extracted from data obtained with a single acoustic transceiver pair. Cornuelle and

Howe (1987) subsequently showed that features which match the ray periodicity (i.e., have the same wavelength) generate large signatures in travel time. The ray paths are somewhat distorted sinusoids in midlatitudes, and so contain higher harmonics. Scales short compared to a double loop length that match these harmonics also generate significant signatures in travel time. This result can be qualitatively understood in the spatial domain (Worcester et al., 1991). Scales short compared to the double loop lengths of the rays affect only the few rays that pass through the feature, leaving the other rays unaffected. This differing sensitivity of the rays gives the fundamental information required to invert for scales short compared to double loop lengths. Features which have wavelengths substantially longer than the ray double loop length (about 50 km in midlatitudes) affect all rays similarly and are indistinguishable from a change in the mean over the entire range between the source and receiver. Travel times are therefore sensitive to ocean features with wavelengths comparable to and shorter than a ray double loop length, as well as to the range-average.

This behavior is most easily understood in the wave number domain, by choosing the model parameters used to represent the ocean sound speed perturbation field to be the complex Fourier coefficients in a spectral expansion. For any specific geometry, the sensitivity of the travel time inverse to various wave numbers can be quantified by plotting the diagonal of the resolution matrix. The resolution matrix gives a particular solution to any linear inverse problem as a weighted average of the true solution (e.g., Aki and Richards, 1980). For the specific case of two moorings separated by 600 km, with a source and five widely separated receivers on each mooring, the diagonal of the resolution matrix for spectral model parameters clearly shows the sensitivity of tomographic measurements to the mean and to harmonics with scales comparable to and shorter than a ray double loop length (Figure A.2). There are obvious spectral gaps for wave numbers between the mean and first harmonics of the ray paths, and again between the first and second harmon-

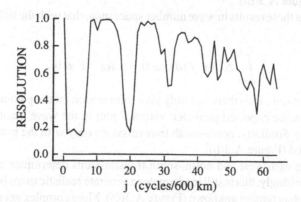

FIGURE A.2. "Transfer function" (the diagonal elements of the resolution matrix) for a 156-ray dataset with an expected variance spectrum (for the lowest baroclinic mode) that is constant for wave numbers $j = 1$ to 3 and decreases as j^{-2} for $j > 3$ (wavelengths smaller than 200 km) (Reproduced from Cornuelle and Howe, 1987.)

ics. Travel times are not sensitive to wave numbers in these regions. The harmonics extend over bands of wave numbers because the eigenrays connecting the source and receiver have a range of double loop lengths.

A.2.3   Horizontal Slice

As was the case for the vertical slice, the key to understanding the horizontal sampling properties of acoustic travel times is to consider the wave number domain, rather than physical space (Cornuelle et al., 1989). For simplicity, consider a two-dimensional ocean consisting of a horizontal slice. Rays are then straight lines connecting sources and receivers (neglecting horizontal refraction, which is usually small due to the small horizontal gradients in the ocean). The sound speed perturbation field $\Delta C(x, y)$ can be expanded in truncated Fourier series in x and y, giving the spectral representation

$$\Delta C(x, y) = \sum_k \sum_l P_{kl} \exp \frac{2\pi i}{L}(kx + ly), \qquad k, l = 0, \pm 1, \ldots, \pm N,$$

where L is the size of the domain and k, l are the wave numbers in x and y, respectively. With this representation of the ocean, the inverse problem is to determine the complex Fourier coefficients P_{kl} from the travel time data. Consider a scenario in which two ships start in the left and right bottom corners of a 1 Mm square and steam northward in parallel, transmitting from west to east every 71 km for a total of 15 transmissions (Figure A.3(a)). The inversion of the 15 travel times leads to an estimate which consists entirely of east–west contours. All of the ray paths give zonal averages, with no information on the longitudinal dependence of the sound speed field. Similarly, transmissions between east-to-west moving ships give meridional averages, and the resulting estimate consists entirely of north–south contours (Figure A.3(b)).

To interpret these results in wave number space, note that with the field expressed as above,

$$\int_0^L \Delta C(x, y)\, dx = 0 \qquad \text{for} \quad k \neq 0.$$

East–west transmissions therefore only give information on the parameters P_{0l}, as can be seen in the expected predicted variance plot in the wave number space of Figure A.3(a). Similarly, north–south transmissions only with the parameters P_{k0} are determined (Figure A.3(b)).

Combining east–west and north–south transmissions determines both P_{0l} and P_{k0}. Not surprisingly, this is still inadequate to generate realistic maps because most of the parameters remain unknown (Figure A.3(c)). More complex geometries with scans at 45° give a distinct improvement by determining the parameters for which $k = l$, but at the cost of excessive ship time (Figure A.3(d)). In all cases, the division between well-determined and poorly determined parameters is simplest in the spectral space.

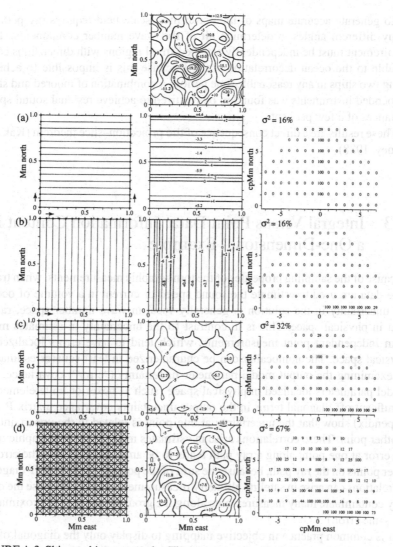

FIGURE A.3. Ship-to-ship tomography. The top center panel is the "true ocean," constructed assuming a horizontally homogeneous and isotropic wave number spectrum with random phases. Energy decreases monotonically with increasing scalar wave number, giving an approximately Gaussian covariance with $1/e$ decay scale of 120 km. (a) W $\rightarrow$ E transmissions between two northward traveling ships (left panel). Inversion of the travel time perturbations produce east–west contours in ΔC (middle) with only a faint relation to the "true ocean." Expected predicted variances in wave number space (right) are 0% (no skill) except for $(k, l) = (0, 1), (0, 2), \ldots, (0, 7)$, which accounts for $\sigma^2 = 16\%$ of the ΔC variance. (b) S $\rightarrow$ N transmissions between two eastward traveling ships. (c) Combined W $\rightarrow$ E and S $\rightarrow$ N transmissions, accounting for 32% of the ΔC variance and giving a slight pattern resemblance to the true ocean. (d) Combined W $\rightarrow$ E, S $\rightarrow$ N, SW $\rightarrow$ NE, and SE $\rightarrow$ NW transmissions, accounting for 67% of the variance and giving some resemblance to the true ocean. (Adapted from Cornuelle et al., 1989.)

To generate accurate maps of the ocean mesoscale field requires ray paths at many different angles to determine all of the wave number components. This requirement must be independently satisfied in all regions with dimensions comparable to the ocean decorrelation scale. Because this is impossible to achieve using two ships in any reasonable time period, a combination of moored and ship-suspended instruments was found to be required to achieve residual sound speed variances of a few per cent.

These results are a direct consequence of the projection–slice theorem (Kak and Slaney, 1988).

A.3   Integral Versus Point Data: Information Content in a One-Dimensional Example

Because of the sampling properties of the tomographic measurements, when travel time data are used to estimate the sound speed or current in a volume of ocean, the uncertainty in the solution is generally local in wave number space, rather than in physical space. This is in contrast to the uncertainty in estimates made from independent point measurements, which tend to have errors localized in physical space. The nonlocality can be characterized in least-squares estimation by examining the output model parameter uncertainty ("error") covariance. For model parameters localized in physical space (such as boxes or finite elements), significant off-diagonal terms in the output uncertainty covariance matrix $\hat{\mathbf{P}}$ (see Appendix) show that the uncertainty at one point is related to the uncertainty at another point. These correlations arise in estimates made from tomographic data; the error at one point along a ray path tends to be anticorrelated with the error at other points on the ray path, because the sum of the points is known. (Off-diagonal correlations can arise in satellite altimetry measurements as well, because the orbit may contaminate many measurements along the ground track with approximately the same error.)

It is common practice in objective mapping to display only the diagonal of the physical space uncertainty covariances, which is what is plotted in the "error map" used in many papers (Bretherton, Davis, and Fandry, 1973). Assimilation methods that insert values at points in the model, such as "nudging" (e.g., Malanotte–Rizzoli and Holland, 1986), sometimes use the local error bars from the objective map to adjust the strength of the data constraint. We will show below that although neglecting the off-diagonal components of the uncertainty covariance of the estimates is benign for point measurements with local covariances, it can be dangerous when the off-diagonal terms are significant, as when tomographic data are used. Doing the data insertion by some form of sequential optimal interpolation (OI), approximating the Kalman filter, is less dangerous, but the treatment of the model parameter uncertainty covariance between steps can still both destroy information and give misleading results when used with nonlocal data.

In order to highlight the contrast between tomographic and point measurements, we have chosen a very simple model problem for pedagogical clarity. We use a one-dimensional, periodic realm of 20 piecewise-linear finite elements, each with identical widths and randomly chosen temperatures, which are assumed to be exactly convertible to sound speeds. The unknowns (model parameters) are the temperatures at each of the points (and by interpolation, the temperature of the entire interval). The model parameters are assumed to be independent, identical, normally distributed random variables with zero mean and independent, equal (unit) variances. The initial model parameter uncertainty covariance matrix (P) is therefore diagonal with 1.0 on the diagonal. The "error map" for this a priori state is a uniform, unit error variance in physical space, which is conveniently the same as the diagonal of the matrix plotted as a function of the location index. If we measure the value of one of these finite elements, say element 5, by sampling it at its center without noise, the output uncertainty variance is now zero for the sampled point (Figure A.4). If we instead measure the integral across all the points (as a representation of a tomographic measurement), the uncertainty variance is instead reduced only slightly at all the points (Figure A.4). This obscures the fact that something very specific has been learned, just as specific as for the point measurement. In fact, the sum of uncertainty variances (the trace of the covariance) has been reduced by the same amount (one unit) in each case, to 19 from 20.

The full uncertainty covariance matrix gives complete information about the character of the remaining uncertainty. The first and fifth columns of the covariance are plotted in Figure A.5 for both point and averaged measurements. (This is the covariance between the error at points 1 or 5 and the error at all the rest of physical space.) The errors at the points missed by the point measurement show no correlation with the errors at their neighbors, while the fifth point (the site of the measurement) shows no error. In contrast, the average shows identical behavior at

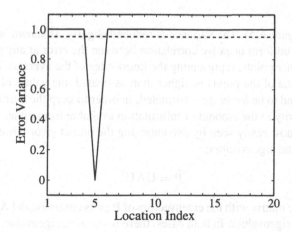

FIGURE A.4. Error variance map (diagonal of the error covariance) from a single, perfect, point measurement at location 5 (solid) or from a single, perfect measurement of the average over 20 points (dashed).

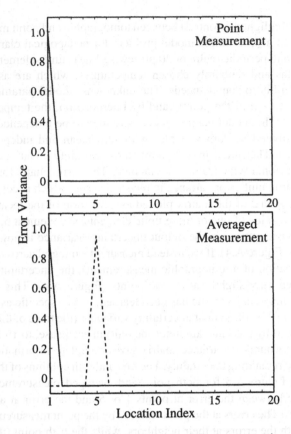

FIGURE A.5. Columns 1 (solid) and 5 (dashed) of the error covariance for the point measurement at location 5 and for the averaged measurement.

all points (Figure A.5). The variance is reduced slightly, as shown in Figure A.4, but there is a uniform negative correlation between the error at any point and the error at all other points, representing the knowledge of the average. That is, if the value at the one of the points is higher than estimated, the values at all the other points will tend to be lower than estimated, in order to keep the average the same.

The similarity in the amount of information available from either type of measurement is most easily seen by decomposing the output error covariance $\hat{\mathbf{P}}$ into eigenvalues and eigenvectors:

$$\hat{\mathbf{P}} = \mathbf{U}\mathbf{\Lambda}\mathbf{U}^T,$$

where $\mathbf{U}$ is the matrix with the eigenvectors of $\hat{\mathbf{P}}$ as its columns, and $\mathbf{\Lambda}$ is the diagonal matrix of eigenvalues. In both cases, there is one zero eigenvalue, representing the component of the model that is known exactly, and 19 eigenvectors with eigenvalue 1. The null space in each case is degenerate, and can be represented by many different sets of basis functions, but the eigenvector that is known is either a spike

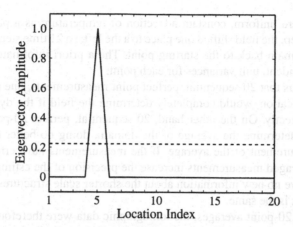

FIGURE A.6. The eigenvector with zero eigenvalue (the single, perfectly known component of the model), for the point measurement at location 5 (solid) and the average measurement (dashed).

at the measured point or a uniform level across all points (Figure A.6). In either case, only 19 unknowns are left to be determined.

Representing the estimate generated from the average as 20 point measurements with equal values and independent error bars with values as in Figure A.4, amounts to dropping the off-diagonal terms in the error covariance. Although the diagonal-ized covariance has the same trace as the exact covariance, and thus has a similar amount of information as measured using the trace, the eigenvectors of the diag-onalized covariance are 20 unknown functions, each with variance 19/20 of the original. This is a considerably different state of knowledge than in the original error covariance, because 20 unknowns remain to be determined, and the slight reduction in their uncertainty is not very useful.

A.4   Integral Versus Point Data: Estimation in a Time-Dependent One-Dimensional Example

Suppose now that the measurements are repeated, and that we wish to combine all the measurements to obtain an improved estimate of the field, assuming that we have a model for the dynamical evolution of the field. The time-dependent least-squares estimation problem can either be solved sequentially (the Kalman filter), or globally, giving identical results when fully optimal methods are used. In the examples to follow, the estimates have been described as the result of sequential estimation, since that is similar to many approximate schemes.

A primitive dynamical example that includes advection demonstrates the effect of error propagation in a data assimilation scheme using nonlocal data. The 20-point domain was retained from the earlier example, but is now assumed to be periodic, with periodicity 20, so that the twenty-first point is the first point. The dynamics were uniform, constant advection of temperature as a passive tracer: every time step, the field shifted one place to a the left, so 21 time steps completely rotate the domain back to the starting point. The a priori information state was again independent, unit variances for each point.

It is obvious that 20 sequential, perfect point measurements, one per time step at a single location, would completely determine the field if the dynamics were modeled perfectly. On the other hand, 20 sequential, perfect 20-point averages would only determine the average of the domain, doing no better than a single perfect measurement of the average. If the measurements have error bars, then repeated averaged measurements increase the precision of the estimate of the average, but give no new information about the shorter scale structures, since every measurement is the same.

Instead of 20-point averages, the tomographic data were therefore taken to be five-point averages, which do not trivially repeat in the periodic domain and which yield some information on smaller scale structures. To avoid singularities due to perfect measurements, both point measurements and averages were assumed to be contaminated by noise. The point measurements were assumed to have an uncertainty variance of 0.1, representing measurement error. Because each tomographic measurement averages five points, the uncertainty variance was set at 0.02, one-fifth of the error variance assumed for the point measurements, keeping the signal-to-noise variance ratio (SNR) equal between the averaged and point measurements. (This is a relatively arbitrary choice, but is simplest for comparisons, because a single measurement of either type produces the same decrease in model uncertainty.)

We used a Kalman filter (see Appendix) to combine the measurements, and compared the performance after all the measurements had been used. In this example, the tomographic measurements show significantly larger point error bars than the

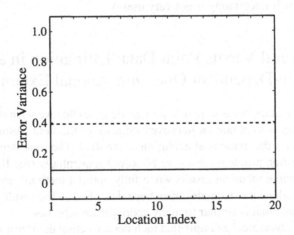

FIGURE A.7. Error variance map (diagonal of output error covariance) with advection after a 20-ste p dynamical inverse, using the point measurements (solid) and the five-point averages (dash d).

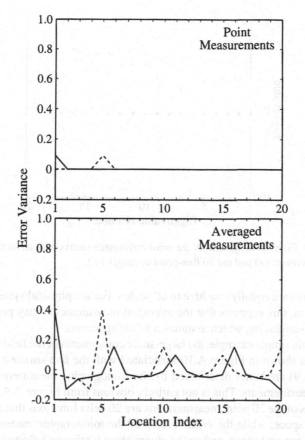

FIGURE A.8. Columns 1 (solid) and 5 (dashed) of the error covariance with advection for the 20 point measurements and for the 20 five-point averages.

point measurements, even though the estimate used optimal error propagation (Figure A.7). Columns 1 and 5 of the output error covariance matrix (Figure A.8) show the contrast in structure between the point measurements and the tomography. The eigenvalue spectra (Figure A.9) show that the averaged data have a varying information content in spectral space. The 20-point mean is well determined (the smallest eigenvalue). The eigenfunctions corresponding to the next two smallest eigenvalues resemble sine and cosine functions with one cycle in the 20-point domain (Figure A.10). The eigenfunctions corresponding to the next two smallest eigenvalues similarly resemble sine and cosine functions with two cycles in the 20-point domain (Figure A.10). The four vectors in the null space, which each average to zero in any five-point domain and are thus completely undetermined by the tomography, appear to be more complex, but correspond to aliased samples from sine and cosine functions that have either one or two cycles over the five-point domain (Figure A.10). In spectral terms, the averaged measurements selectively determine the large-scale components better than the short-scale components, while the point

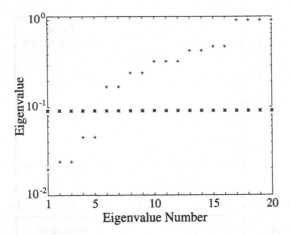

FIGURE A.9. Eigenvalue spectra for the error covariance matrix with advection for the 20 point measurements (∗) and the 20 five-point averages (+).

measurements are equally sensitive to all scales. For geophysical systems with red signal spectra, this suggests that the averaged measurements may perform better than in this simulation, which assumes a white spectrum.

Even in this simple example, the large-scale components of the field (i.e., the five eigenvectors shown in Figure A.10 associated with the five smallest eigenvalues in Figure A.9) are better determined by the tomographic measurements than by the point measurements. This is not entirely obvious from Figure A.9, because the eigenvectors of the 20 point measurements are 20 delta functions that are localized in physical space, while the eigenvectors of the tomographic measurements are localized in spectral space, and so the eigenvalues (variances) plotted in the figure do not correspond to similar eigenvectors for the two cases. Eigenvector 1 for the 20 five-point averages is related to the mean, for example, while eigenvector 1 for the 20 point measurements is a delta function at location 1. The variance of the mean deduced from the 20 point measurements is simply one-twentieth of the 20 identical variances (eigenvalues) of the 20 eigenvectors. Because the eigenvectors are normalized to have unit length, the elements of eigenvector 1 for the 20 five-point averages all have magnitude $1/\sqrt{20}$, as can be seen in Figure A.10. The variance of the mean is then one-twentieth of the variance (eigenvalue) of the eigenvector. The ratios of the eigenvalues in Figure A.9 therefore accurately reflect the ratios of the error variances of the estimates made using tomographic measurements and point measurements, showing that the large-scale components are better determined by the tomographic data. (Another way of looking at this is to note that because this simple example is homogeneous, sines and cosines are also eigenvectors of the point measurement covariance.)

A somewhat different question is how well the tomographic and point measurements resolve the detailed spatial structure of the field, rather than just the large-scale components. The tomographic measurements were seen in Figure A.7 to have significantly larger point error bars than the point measurements for the

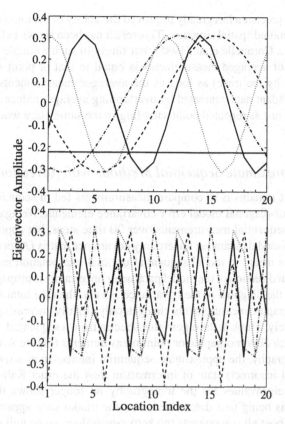

FIGURE A.10. The five best-determined eigenvectors (top) and the four null space vectors (bottom), for 20 five-point averaged measurements with advection.

case with 20 measurements. In that case, however, the tomographic measurements had more null space vectors than the point measurements. To give the tomographic and regular measurements similar numbers of null space vectors, we repeated the simulations using only 16 data in each case. The null space for the point measurements has four elements, representing the four points not measured, while the null space for the tomography remains the same. The trace of the output error covariances (the total uncertainty variance after the inverses) were 9.2 and 5.5 for the averaged and point measurements, respectively. The point measurements thus do better in resolving the detailed spatial structure of the field than the averaged measurements, when the unknown field has a white spectrum and when both types of data have equal SNR. The averaged measurements are most sensitive to the larger scales, as discussed above, and have to be differenced in order to resolve finer scales. The difference of two large numbers is easily contaminated by random noise. If the calculations are repeated giving the tomography data 0.1 of their original variances (tomographic SNR = 10 time point measurement SNR), the trace of the output error covariance for the tomography is now 5.1, so the greater measurement precision has greatly improved the ability of the tomographic data to resolve the detailed spatial structure. This result has been shown before in number of places (e.g., Cornuelle et al., 1985), but rarely in such a simple example. The performance of averaged measurements is equal to that of point measurements (as measured by the trace) as long as the averaged measurements do not overlap. The redundant data generated by overlapping averages reduce the calculated performance, just as repeated point sampling in the same place would.

A.4.1   Approximate Sequential Methods with Advection

The issue that remains is to compare measurements fed in sequentially without keeping the off-diagonal model error covariance elements. We again use the example of 20 sequential measurements over 20 time steps, but approximating the forecast of the model parameter uncertainty covariance matrix (Appendix, equation (A.7)). This is meant to model sequential optimal interpolation methods, where only a simplified version of the model parameter uncertainty is propagated between steps. If only the diagonal of the covariance is kept, then the total expected error for the tomographic measurements changes only slightly, increasing by about 3

Unfortunately, a look at the eigenvalue spectra from a sequential, diagonal-only estimation with 20 tomographic or point measurements (Figure A.11) shows that for the tomography, the approximate sequential interpolation arrived at a vastly different (and incorrect) state of information than the exact Kalman filter. The spectrum of eigenvalues for the tomography no longer shows the large-scale components as being best determined, and the model state apparently includes information about all components (no zero eigenvalues, so no null space). This is in contrast to the point measurements, whose eigenvalue spectrum is unchanged.

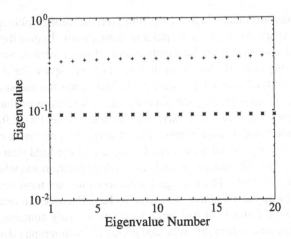

FIGURE A.11. Same as Figure A.9, but for a simulation in which only the diagonal of the covariance matrix was kept.

Because of the approximate error propagation, the error covariance is no longer a good figure of merit, and the true performance of the simplified method can best be evaluated by Monte Carlo methods, simulating an ensemble of true fields and looking at the error in the reconstruction. We only wish to point out that the diagonal-only method remains optimal for the point measurements, while becoming severely suboptimal for the tomographic measurements, but in a subtle way that could easily be overlooked. This contrast is heightened by the trivial dynamics chosen for the simulations. Realistic dynamics, such as quasi-geostrophic flow in three dimensions, generally creates nonlocal covariances, even from point sampling, so that the sequential optimal interpolation would degrade the point measurements somewhat. On the other hand, for short time scales and normal advection velocities, the point measurement information will remain much more local than tomographic information, and so is more compatible with local approximations. Conversely, if the dynamical model is built in spectral space, so the horizontal basis functions are sines and cosines, then the tomographic data is much more local than point measurements, which are sensitive to all scales.

Most modern data assimilation methods do not completely ignore off-diagonal terms in the model parameter uncertainty covariance matrix, however, even for point measurements. It is therefore natural to ask how well other possible approximations to the uncertainty covariance matrix perform. Perhaps the simplest class of approximations are ones in which varying numbers of diagonal bands of off-diagonal elements are retained, while the remaining elements are set to zero. Plotting the eigenvalue spectra as a function of the number of bands retained (Figure A.12) shows that retaining one off-diagonal band, in addition to the diagonal

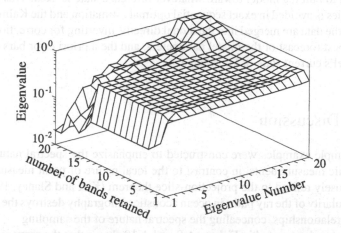

FIGURE A.12. Eigenvalue spectra for the error covariance matrix with advection for the 20 five-point averages, as a function of the number of bands of off-diagonal elements retained in the model parameter uncertainty covariance matrix. The spectrum obtained retaining only one band (i.e., the diagonal) is the same as that in Figure A.11; the spectrum obtained retaining 20 bands (i.e., the full matrix) is the same as that in Figure A.9.

elements, results in the reduction of a single eigenvalue, corresponding to the mean. Little further change in the spectra is evident as additional off-diagonal bands are retained, until 15 off-diagonal bands are included. At that point the spectra begin to resemble the spectrum obtained when the full matrix is used. For the simple example considered here, retaining additional off-diagonal bands of the uncertainty covariance matrix is therefore not a particularly effective approximation, as nearly the complete uncertainty covariance matrix needs to be retained before the results are similar to those obtained using the full matrix.

The decomposition of the error covariance into eigenvectors suggests a more natural approximation for sequential assimilation, however, in which only the components of the model error covariance with large eigenvalues are propagated by the model. In the case of a single measurement, the savings are small, because 19 out of 20 vectors need to be propagated, but with more complete observations, the savings could be larger.

A.4.2   Separating the Inverse from the Assimilation

Even the approximate method used in the previous example kept the inverse as part of the update of the model. Some older assimilation methods invert the measurements and then blend in the results as pseudo-point measurements with error bars. This approach is impossible when using the averaged measurements, because the uncertainty of the output estimate is not local, and so the pointwise error bars cannot express the infinite (but correlated) uncertainty imposed on the solution by the elements of the null space. Even if the data are inverted outside the model, it is necessary to use the model state as the reference; otherwise the inversion procedure will tend to pull the model toward whatever reference state is used. This problem of infinities is avoided in exact sequential optimal estimation and the Kalman filter, because the data are merged into the model directly, inverting for corrections to the current best forecast of the model parameters, and the a priori error bars describe the model's current state of knowledge.

A.5   Discussion

These simple examples were constructed to emphasize the spectral nature of tomographic measurements, in contrast to the local nature of point measurements. This is losely related to the projection-slice theorem (Kak and Slaney, 1988), but the irregularity of the ray paths in ocean acoustic tomography destroys the simplest spectral relationships, concealing the spectral nature of the sampling.

The example reported in Figures A.9 and A.10 shows that the error covariance matrix of the averaged measurements has sines and cosines as eigenvectors, while the error covariance matrix of the point measurements is diagonal with delta functions as one set of eigenvectors. For an unknown field with a white spectrum, and data with equal SNRs, nonoverlapping averaged measurements increase our knowledge of the unknown field by the same amount as the same number of point measurements, but the spectral content of that knowledge is very different. Because the averaged measurements determine the lower wave numbers better than the higher wave numbers, they have advantages if the spectrum of the unknown field is red. Determining high wave number information from the averaged measurements is more difficult, unless the measurement errors are sufficiently small to make differencing of the integral measurements practical. The relative utility of tomographic measurements and point measurements thus depends strongly on the goal of the measurement program.

The nonlocal nature of the averaged measurements also makes it difficult to use approximations to the Kalman filter in dynamical models with local parameterization. Conversely, the averaged measurements can be used efficiently by an approximate Kalman filter based on spectral functions.

Acknowledgments. This work was supported by the Office of Naval Research (ONR Contracts N00014-93-1-0461 and N00014-94-1-0573) and by the Strategic Environmental Research and Development Program through the Advanced Research Projects Agency (ARPA Grant MDA972-93-1-0003).

A.6   Appendix

The form of least-squares estimation used here assumes that the expected value of the model parameter vector has been removed, so $\langle m \rangle = 0$, and that an initial guess exists for the covariance of the uncertainty around the expected value, $\langle mm^T \rangle = P$. The data are related to the model parameter vector by a linear relation

$$d = Gm + n, \tag{A.1}$$

where n is the random noise contaminating the measurements. Any known expected value of the noise is assumed to have been removed, so $\langle n \rangle = 0$, and the noise is assumed to have covariance $\langle nn^T \rangle = N$ and to be uncorrelated with the model parameters. This relation can be inverted to obtain an estimate of the model parameters

$$\hat{m} = PG^T (GPG^T + N)^{-1} d \tag{A.2}$$

and the expected uncertainty in this estimate is

$$\hat{P} = P - PG^T (GPG^T + N)^{-1} GP. \tag{A.3}$$

If dynamics are available to forecast the model parameter vector between time steps, so that

$$m_{t+1} = Am_t + q, \tag{A.4}$$

where A is the transition matrix, and q is the uncertainty in the forecast due to errors in the dynamics (with zero mean and uncertainty covariance $Q = \langle qq^T \rangle$).

The Kalman filter performs a sequential cycle, correcting the starting guess by inverting the differences between the observations and the predicted data

$$\hat{\mathbf{m}}_t = \mathbf{m}_t + \mathbf{P}_t \mathbf{G}^T (\mathbf{G}\mathbf{P}_t\mathbf{G}^T + \mathbf{N})^{-1}(\mathbf{d}_t - \mathbf{G}\mathbf{m}_t), \qquad \text{(A.5a)}$$

$$\hat{\mathbf{P}}_t = \mathbf{P}_t - \mathbf{P}_t\mathbf{G}^T(\mathbf{G}\mathbf{P}_t\mathbf{G}^T + \mathbf{N})^{-1}\mathbf{G}\mathbf{P}_t, \qquad \text{(A.5b)}$$

and forecasting the estimate and covariance to the start of the next step

$$\mathbf{m}_{t+1} = \mathbf{A}\hat{\mathbf{m}}_t, \qquad \text{(A.6a)}$$

$$\mathbf{P}_{t+1} = \mathbf{A}\hat{\mathbf{P}}_t\mathbf{A}^T + \mathbf{Q}. \qquad \text{(A.6b)}$$

This cycle then repeats.

A.7   References

[AK80]   K. Aki and P. Richards, 1980. *Quantitative Seismology, Theory and Methods*, 2 Vols. W.H. Freeman, San Francisco.

[B+79]   F.P. Bretherton, R.E. Davis, and C.B. Fandry, 1976. A technique for objective analysis and design of oceanographic experiments applied to MODE-73. *Deep Sea Res.*, 23:559–582.

[CD87]   C.-S. Chiu and Y. Desaubies, 1987. A planetary wave analysis using the acoustic and conventional arrays in the 1981 Ocean Tomography Experiment. *J. Phys. Oceanogr.* 17:1270–1287.

[C+87]   C.-S. Chiu, J.F. Lynch, and O.M. Johannessen, 1987. Tomographic resolution of mesoscale eddies in the marginal ice zone: A preliminary study. *J. Geophys. Res.*, 92:6886–6902.

[Co90]   B.D. Cornuelle, 1990. Practical aspects of ocean acoustic tomography. In: Y. Desaubies, A. Tarantola, and J. Zinn-Justin (eds.), *Oceanographic and Geophysical Tomography: Proc. 50th Les Houches Ecole d'Ete de Physique Theorique and NATO ASI*. Elsevier Science Publishers, Amsterdam, pp. 441–463.

[CH87]   B.D. Cornuelle and B.M. Howe, 1987. High spatial resolution in vertical slice ocean acoustic tomography. *J. Geophys. Res.*, 92:11,680–11,692.

[C+89]   B.D. Cornuelle, W.H. Munk, and P.F. Worcester, 1989. Ocean acoustic tomography from ships. *J. Geophys. Res.*, 94:6232–6250.

[C+93]   B.D. Cornuelle, P.F. Worcester, J.A. Hildebrand, W.S. Hodgkiss Jr., T.F. Duda, J. Boyd, B.M. Howe, J.A. Mercer, and R.C. Spindel, 1993. Ocean acoustic tomography at 1000 km range using wavefronts measured with a large-aperture vertical array. *J. Geophys. Res.*, 98:16,365–16,377.

[C+85]   B.D. Cornuelle, C. Wunsch, D. Behringer, T.G. Birdsall, M.G. Brown, R. Heinmiller, R.A. Knox, K. Metzger, W.H. Munk, J.L. Spiesberger, R.C.

Spindel, D.C. Webb, and P.F. Worcester, 1985. Tomographic maps of the ocean mesoscale, 1: Pure acoustics. *J. Phys. Oceanogr.*, **15**:133–152.

[FM95]   I. Fukumori and P. Malanotte-Rizzoli, 1995. An approximate Kalman filter for ocean data assimilation: An example with an idealized Gulf Stream model. *J. Geophys. Res.*, **100**:6777–6793.

[H⁺87]   B.M. Howe, P.F. Worcester, and R.C. Spindel, 1987. Ocean acoustic tomography: Meso-scale velocity. *J. Geophys. Res.*, **92**:3785–3805.

[KS88]   A.C. Kak, and M. Slaney, 1988. *Principles of Computerized Tomographic Imaging*. IEEE Press, New York.

[MH86]   P. Malanotte-Rizzoli and W.R. Holland, 1986. Data constraints applied to models of the ocean general circulation, Part I: The steady case. *J. Phys. Oceanogr.*, **16**:1665–1687.

[M⁺95]   W. Munk, P.F. Worcester, and C. Wunsch, 1995. *Ocean Acoustic Tomography*. Cambridge University Press, Cambridge.

[MW79]   W. Munk and C. Wunsch, 1979. Ocean acoustic tomography: A scheme for large scale monitoring. *Deep-Sea Res.*, **26**:123–161.

[MW82]   W. Munk and C. Wunsch, 1982. Up/down resolution in ocean acoustic tomography. *Deep-Sea Res.*, **29**:1415–1436.

[OTG82] Ocean Tomography Group, 1982. A demonstration of ocean acoustic tomography. *Nature*, **299**:121–125.

[JW86]   J. Schröter and C. Wunsch, 1986. Solution of nonlinear finite difference ocean models by optimization methods with sensitivity and observational strategy analysis. *J. Phys. Oceanogr.*, **16**:1855–1874.

[Sh89]   J. Sheinbaum, 1989. Assimilation of Oceanographic Data in Numerical Models. Ph.D. Thesis, University of Oxford, Oxford, England, 156 pp.

[SM91]   J.L. Spiesberger and K. Metzger Jr., 1991. Basin-scale tomography: A new tool for studying weather and climate. *J. Geophys. Res.*, **96**: 4869–4889.

[W⁺91]   P.F. Worcester, B.D. Cornuelle, and R.C. Spindel, 1991. A review of ocean acoustic tomography: 1987-1990. *Reviews of Geophysics, Supplement, U.S. National Report to the International Union of Geodesy and Geophysics 1987–1990*, pp. 557–570.

[Wu90]   C. Wunsch, 1990. Using data with models: Ill-posed problems. In Y. Desaubies, A. Tarantola, and J. Zinn-Justin, (eds.), *Oceanographic and geophysical tomography: Proc. 50th Les Houches Ecole d'Ete de Physique Theorique and NATO ASI*. Elsevier Science, Amsterdam, pp. 203–248.

Index

Contributors

David Chapman
Defence Research Establishment Atlantic, Dartmouth, NS, B2Y 3Z7, CANADA
Email: dave.chapman@drea.dnd.ca

N. Ross Chapman
School of Earth and Ocean Sciences, University of Victoria, Victoria, BC, V8W 3P6, CANADA
Email: chapman@unic.ca

Stan E. Dosso
School of Earth and Ocean Sciences, University of Victoria, Victoria, BC, V8W 3P6, CANADA
Email: sdosso@uvic.ca

B. Duchêne
DRE - LSS, SUPPLEC, Plateau de Moulon, F-91192 Gif-sur-Yvette, FRANCE
Email: duchene@supelec.fr

Zoi-Heleni Michalopoulou
New Jersey Institute of Technology, Newark, NJ 07102, USA
Email: elmich@aqua.njit.edu

Peter N. Mikhalevsky
Science Applications International Corporation, McLean, VA 22102, USA
Email: peter@osg.saic.com

James H. Miller
Department of Ocean Engineering, University of Rhode Island, Narragansett, RI 02882, USA
Email: miller@trident.oce.uri.edu

Alex Tolstoy
ATolstoy Sciences, Annandale, VA 22003, USA
Email: atolstoy@ieee.org

Peter Worcester
Scripps Institution of Oceanography, University of California at San
Diego, La Jolla, CA 92093, USA
Email: pworcester@ucsd.edu

Valery Yakhno
Sobolev Institute of Mathematics, Russian Academy of Sciences,
Koptyug prospekt 4, 630090 Novosibirsk, RUSSIA
Email: yakhno@math.nsc.ru